HELEN GORDON's books include a novel, *Landfall*, and *Being a Writer*, a compendium written with Travis Elborough. She is married to an earth scientist and lives mostly in the Holocene.

Praise for *Notes from Deep Time*

'Astounding ... To call this a "history" does not do justice to Helen Gordon's ambition. Her adventures in the deep time of Earth hark all the way back to its beginnings as a barren ocean planet, 4.4 billion years ago, while keeping one foot firmly planted in the depleted and desertified plaything we're left with today ... *Notes from Deep Time* sidesteps the maundering and finger-wagging that comes with much Anthropocene thinking and shows us how much sheer intellectual and poetical entertainment there is to be had in the idea'
Simon Ings, *Daily Telegraph*

'If there were ever a good time to think about deep time, it's now ... A whirlwind tour of our planet's deep past and far future ... succeeds in grounding our existence firmly in the context of geological time'
Alexandra Witze, *Nature*

'A book as multi-layered as the deep-time planet itself'
Sara Wheeler, author of *Terra Incognita*

'Sublime ... a fascinating and thrilling descent into time, human in scale but full of moments of vertiginous wonder'
Jon Day, author of *Homing*

NOTES FROM DEEP TIME

A Journey Through Our Past and Future Worlds

HELEN GORDON

PROFILE BOOKS

This paperback edition first published in 2022

First published in Great Britain in 2021 by
Profile Books Ltd
29 Cloth Fair
London
EC1A 7JQ

www.profilebooks.com

Chapter 2: Box 48903C16: An earlier version of the Paris material (though not the Danish sections) appeared on www.1843magazine.com; Chapter 8: The Fiery Fields: An earlier version of this chapter appeared in *1843* magazine; Chapter 12: Colouring Deep Time: An earlier version of this chapter appeared on www.wired.co.uk; Chapter 13: Urban Geology: An earlier version of this chapter appeared in *1843* magazine; Chapter 14: In Search of the Anthropocene: An earlier version of this chapter appeared in *1843* magazine; Chapter 15: 'This is Not a Place of Honor': An earlier version of the Onkalo material appeared in *Wired* magazine; an earlier version of the Cigeo material appeared on https://mosaicscience.com/

3 5 7 9 10 8 6 4 2

Typeset in Sabon by MacGuru Ltd
Printed and bound in Great Britain by
CPI Group (UK) Ltd, Croydon, CR0 4YY

A CIP catalogue record for this book is available from the British Library.

ISBN 978 1 78816 164 0
eISBN 978 1 78283 504 2

For
Jonny

CONTENTS

Eon	Era	Period	Millions of years before present	Epoch
Phanerozoic	Cenozoic	Quaternary	2.58	HOLOCENE PLEISTOCENE
		Neogene	23	
		Palaeogene	66	
	Mesozoic	Cretaceous	145	
		Jurassic	201	
		Triassic	252	
	Palaeozoic	Permian	299	
		Carboniferous	359	
		Devonian	419	
		Silurian	444	
		Ordovician	485	
		Cambrian	541	
Pre-Cambrian		*Proterozoic Eon*	2,500	
		Archean Eon	4,000	
		Hadean	4,600	

1

DEEP TIME ON
CAMBRIDGE HEATH ROAD

'Ten thousand years is nothing,' the geologist told me. 'Ten thousand years ago is basically the present.'

Ten thousand years ago Britain was still a peninsula connected to the continental mainland. In America meltwater from retreating ice sheets was filling up the great lakes: Superior, Michigan, Huron, Erie, Ontario. Worldwide, the human population numbered only a few million. If ten thousand years is basically nothing, then it's a nothing that encompasses the entirety of recorded human history, from the development of writing to space travel and the atom bomb.

Geologists, I was beginning to realise, see the world a little differently from other people. It comes from living half inside what we might call human time and half inside another larger, weirder scale – that of deep time. If human time is measured in seconds and minutes, hours and years, then deep time deals with hundreds of thousands of years, with the millions and the billions. Thinking about it engenders a sort of temporal vertigo. To live in deep time is to take the long view, which means getting your head into a somewhat different place. In deep time it is not just what happened last week or last year or last decade that matters – it's also what happened a million, 50 million, 500 million years ago. It's about the ways in which the succession of events across those millions of years can be said to explain why you're here right now in this particular place, in this particular moment.

*

Some time ago I became a little obsessed with the bright white chalk of the North Downs, the long ridge of hills that push up out of suburban south London. It was late January. The previous year a longish relationship had ended, and on New Year's Day what might have become a new relationship had also finished. The explanation from the man involved had been couched, somewhat confusingly, in a reference to the ending of J. M. Coetzee's *Disgrace* – a novel that I'd read but which seemed, and continues to seem, to me, to have little bearing on that particular romantic situation. Feeling the need for distraction and a change of scenery, I bought a train ticket.

Travelling south out of London, it's when you reach the North Downs that you begin for the first time to get a sense of separation from the city. Sitting on the broad back of a fallen oak tree, looking out across a bare, flinty field towards far-off grey and silver towers, you might begin to get some perspective on things – distance, at least.

After lunch I followed a path along the ridge, sticky brown mud slipping over soft white rock. Somewhere between the commuter towns of Coulsdon and Caterham, I came across an information board conveying several simple but confounding facts. That the ground I was walking on was the remains of a long-vanished, prehistoric ocean. That this ocean had disappeared shortly after the end of the age of the dinosaurs. That whenever you are standing on chalk, you are standing where the sea used to be.

Wanting to know more, I visited the Natural History Museum in South Kensington and smaller, local museums where rows of specimens were displayed in dusty cabinets with narrow labels written out on long-defunct typewriters. I read introductions to geology and talked to sedimentologists and stratigraphers and palaeontologists. I joined field trips to quarries and exposed cliff faces, and learned that the history of deep time is written in the rocks all around and underneath us. In a lump of chalk I found a milky-grey spherical sea sponge the size of my smallest fingernail, its surface pricked with countless tiny holes. I read that some scientists believe that sponges were the first animal group to branch off the evolutionary tree from our common ancestor, making them the sister group of all other animals.[1]

*

Several years after that trip to the North Downs, I stood one summer afternoon peering through the chain-link fence of a building site on Cambridge Heath Road in east London. It was a little after 5 p.m., and the workmen had gone for the day, leaving behind just a solitary excavator looking, as excavators do, a little beastish with its drooping, angular neck and large metal jaw resting on top of a mound of dark earth. It was the hole the excavator had dug that I was interested in.

Walking around London, we know, if we care to think about it, that beneath our feet are many layers of rock, most of which have never been seen by human eyes because there were no eyes to see them when they were first formed, or through the long story of them being buried and hidden and lost. Anyone with a thirst to tread on *terra incognita* might as well dig downwards in their own back garden as travel to the middle of Antarctica. Geologists learn to read these layers and from them construct a story about the past. Each layer represents a former world that came into being, existed for thousands or millions of years and then vanished, compressed into a layer of rock.

'Most humans are chronophobes,' the geologist Marcia Bjornerud has written.[2] 'We worry about where the time has gone, whether we're spending it wisely, how much of it we have left. Geology puts things in temporal perspective.' An excavation in the middle of the city is a portal to the past, a space to look backwards and recalibrate. For the last month I'd been searching for such a site. Then Jonny, my husband, messaged me from his office. On the train into Liverpool Street he'd spotted the excavation on Cambridge Heath.

The sides of the hole showed three distinct layers of earth and rock set neatly on top of one another like a layered pink, white and yellow angel cake. The preciseness of the layers gave the appearance of an illustrative diagram in a geology textbook. The top layer was about a metre deep. A confusion of pale greyish brown earth filled with fragments of broken orange and dusty pink brickwork, lumps of black tarmac and nobbly clumps of cement. This is what geologists call 'made ground' – in a city it's the stuff that will have been constantly rehashed, recycled and added to by successive generations. Made ground is human history, like the artefacts in the V&A Museum of Childhood across the road, where, when I was a child, my parents sometimes took me on a wet Saturday afternoon. If – I suppose I should write 'when' – we disappear from this planet, the made ground is one of the things we'll leave behind. A footprint. A sign saying, 'We Were Here'.

The layer below the made ground was damp sand and gravel

the colour of yellow sponge cake soaked in tea. We know that this layer is older than the made ground because of the work of a seventeenth-century Danish doctor: Niels Stensen, better known as Nicolas Steno. Studying the formation of sedimentary rocks such as those beneath London – i.e., rocks formed (often underwater) from the deposition of tiny fragments of older rocks or fossil remains, or from chemical processes such as the evaporation of seawater – Steno observed that, for a new layer of sediment to accrete, there must already exist a firm layer for it to settle on. Older sedimentary rock layers therefore underlie newer ones.

The damp sand and gravel weren't very far below the surface – only a metre or so down from where buses lumbered through the rush-hour traffic and signs outside the bar under the railway arches advertised '£2.50 tequila and Jager shots ALL DAY LONG!' – but there was no evidence of any human presence in this layer. As the workmen on the building site dug down, they had travelled beyond the cosy familiarity of human time into the world of deep time. What the evolutionary biologist Stephen Jay Gould called geology's 'most distinctive and transforming contribution to human thought'.[3]

*

The layer of sand and gravel in the excavation on Cambridge Heath Road was deposited about 2 million years ago during a unit of time known as the Pleistocene Epoch, when the Thames flowed through what is now Bethnal Green, following a course somewhat north of its present position. Staring down at the damp sand, I tried to think about 2 million years. It's a number that's easy enough to write down but difficult to really comprehend.

'The great challenge is to get people to really understand

the immense amounts of time we're dealing with,' a friend who lectures in geology told me. A report commissioned by the Natural History Museum calls deep time is 'foundational to our full understanding of life's origin and diversification, it is a critical concept for understanding geology, physics, and astrophysics.'[4] We need to grapple with deep time if we want to make sense of the world around us, the long march of evolution, the rapidly multiplying challenges of climate change that threaten life as we think we know it. Without deep time we cannot begin to answer the questions 'Why am I here?', 'Where have I come from?' and 'Where am I going?'

In the world of deep time, the 2 million years since the sandy layer was deposited is not very long. The earliest vertebrates lived more than 500 million years ago. Photosynthesis goes back at least 3 billion. And faced with all these millions and billions, the brain rebels, refuses to engage fully. Perhaps this is a psychological defence mechanism. In the UK average life-expectancy is 81 years. In the US it is a little less – 79 years; in Japan it's 84 years – a little more.[5] We find it difficult to conceptualise much beyond five generations: two behind and two in front. As the Scottish scientist and mathematician John Playfair wrote about geological time in 1802, 'how much farther reason may sometimes go than imagination may venture to follow'.[6]

In the Museum of Childhood there is a doll's house from seventeenth-century Holland, built around the same time that Steno was formulating his work on sedimentary rocks. There is a miniature Dutch-style kitchen complete with Delft tiles, pewter plates, intricate jelly moulds. It was probably created not for a child but for some wealthy woman.

Who was she? That information has not been recorded. The three centuries that have passed are easily long enough for a woman's name to disappear. From the perspective of deep time, however, the unknown seventeenth-century Dutch woman and I exist essentially in the same moment along with all the rest of human history.

*

Below the sand and gravel the earth changed again, and the next layer I recognised as the London Clay. Thickly sticky, a morose dark brown almost purple in places, like many rocks it was formed through geological or 'deep time' processes (in this case, sedimentation and burial) operating at speeds so slow as to be invisible to humans. To watch the creation of around a metre of the London Clay you'd need, in addition to a time machine, a massively powerful time-lapse camera trained for several hundreds of thousands of years on the sediment collecting on a prehistoric seabed. In deep time things happen very, very slowly, but they happen over long enough periods to have huge effects. Here a new rock formation is laid down, there a section of the sea floor is lifted up to become the top of a mountain. The top of Mount Everest was once a seafloor.

The presence of the London Clay in Bethnal Green indicates that around 55 million years ago this area was covered by a warm tropical sea. Were you able to travel there, you would find, somewhere near by, a lush green shoreline with a climate similar to that of present-day Indonesia. A place where *Hyracotherium*, a fox-sized ancestor of the horse, grazed between *Nypa* mangrove palms and waxy-petalled magnolia.

'The most interesting thing about going on a field trip with geologists is their imagination,' the Geological Society of London librarian and poet Michael McKimm told me when I visited the society's headquarters in London's Piccadilly.[7] 'You're all standing on a beach and what they are trying to imagine is why a certain rock structure exists, what happened in the past to get to that point.' As the eminent nineteenth-century geologist Charles Lyell put it: 'we may restore in imagination the appearance of the ancient continents which have passed away.'[8]

Among the sciences it is something of a curiosity: a discipline that involves building worlds in your mind and presenting them to others using descriptive language. For a writer with a background in literary publishing – a sphere where people who perhaps spend too much time thinking about language tend to congregate – geology had an immediate appeal. I recognise something of this impulse in the words of the American writer John McPhee – often credited with the first use of the term 'deep time', in his book *Basin and Range* (1981) – as he describes his early encounters with geology, reflecting that 'There seemed, indeed, to be more than a little of the humanities in the subject: Geologists communicated in English; and they could name things in a manner that sent shivers through the bones.'[9] *Batholiths*, McPhee wrote. *Xenoliths. Desert pavements. The slip-face of a barchan dune.*

In the introduction to one well-known geology textbook – *Geological History of Britain and Ireland*, by Nigel Woodcock and Rob Strachan – the authors write: 'Philosophers of science have struggled to characterize the way a geologist works and thinks. Having identified physics as the quintessential science, they have typically measured other sciences against its supposed objectivity, predictability and precision. Geology has therefore been viewed merely as a derivative and imprecise form of physics.'[10] In the scientific pecking order, theoretical physicists look down on the experimental physicists, who look down on geologists. 'Who do geologists look down on?' I asked my lecturer friend. 'Geographers,' he said.

'Geology,' Woodcock and Strachan write, 'has an essential

historical dimension, which distinguishes it from pure physics, chemistry or biology. The geological record is inevitably complex and incomplete, and deciphering it requires an interpretative reasoning similar to that applied to human history.'[11]

Or as one geologist put it to me, the science requires 'grey data skill sets'. The ability to piece together a story from incomplete, missing or fragmentary data. To use imagination to complete a half-formed picture. Or, as another said: 'It's Sherlock Holmes stuff, basically.'

A few years ago the Geological Society held a celebration of poetry and geology. 'As far as I'm aware we are the only science society that has held a poetry day really driven by its members,' McKimm told me. The then president, Bryan Lovell, read an extract from Alfred Tennyson's 'In Memoriam A. H. H.'. Completed in 1849, just forty years after the establishment of the Society – which is the world's oldest national geological group – the lines reflect the shifting world of deep time as newly revealed by the Victorian geologists:

The hills are shadows, and they flow
From form to form, and nothing stands;
They melt like mist, the solid lands,
Like clouds they shape themselves and go.[12]

'Poets and geologists have a common cause,' Lovell told the assembled crowd. 'A search for words to help us to understand what we do.'

*

Several years after my first visit, I went back to Cambridge Heath Road. Where the hole had been was a six-storey hotel.

Its bar had bare, over-sized light bulbs and exposed pipes. There were Massage Mondays and Nespresso machines in every room. I drank a glass of ginger ale while a Spanish couple scrolled through their phones and members of the European Investment Bank's sports and cultural club milled around shouldering matching sports bags. Across the road, a group of schoolchildren in bright yellow tabards lined up in crocodile formation, two by two, in front of the museum entrance.

Beneath us were two basement levels and beneath that the world of the London Clay. From the London Clay layer continue to the next layer and back another 30 million years and you would find a vast ocean filled with long-jawed ichthyosaurs, flippered plesiosaurs, and razor-toothed, blunt-nosed sharks. Another 50 million years and you would be on dry land: a steep-sided mountain, foothills fringed with tropical forests, lakes and marshes frequented by ancestral crocodiles basking, one assumes, on glistening prehistoric mud. Worlds beneath worlds beneath worlds. Millions and millions of years stacked on top of one another like a deck of playing cards.

Were we able to watch all of deep time unfolding – another time-lapse film – we'd see hot dry deserts turn into lush jungles, rise up as craggy mountains, wear down to a low line of hills. A shifting, sliding cartography. In deep time everything is provisional. Bones become rock. Sands become mountains. Oceans become cities.

And being conscious of the immense span of time necessary to encompass all of these miraculous changes reminds us again that our own allotted span is shockingly brief – as individuals, as a species. A friend of mine once took a weekend pottery course and fashioned a terrible, lumpy brown vase. 'Just think,' the tutor brightly told departing students, 'you've all made something that will probably outlast you.' My friend, staring at the misshapen clay, was horrified. *This* was what everything would amount to? The impersonal, eroding sweep of the ages challenges our instinct for memorialisation – the desire to frame photographs and certificates, to put up gravestones, to append our name (if we have the spare cash) to a gallery wing or lecture theatre, to autograph motorway underpasses and the doors of public toilets – our need to ask, what will we leave behind? What will survive us?

*

Like finding a fossil sponge nestling in a seemingly amorphous lump of chalk, I retain a persistent memory of an event from my childhood that may or may not have taken place. The uncertainty is because memory is notoriously unreliable and no one else was present to witness what happened, and because it occurred when I was so young that the memory appears to me free-floating, chronologically unbound and therefore suspect.

I was walking with my parents and brothers in the Firehills, the gorse-covered cliff tops near Hastings on the south coast. Running ahead, coming to a fork in the sandy path, I took the right-hand way towards the cliff edge, ignoring a 'Footpath Closed' sign. As I've told and retold the story to myself, the path led up a slight incline that prevented me from seeing ahead to where, the other side of the rise, what must have been a recent cliff fall had taken out the path. I have a memory of the world suddenly expanding and opening out. The broad sweep of sunlit cliffs, the warm coconut scent of the cadmium-yellow gorse and the far-below shining sea. The few steps between myself and the cliff edge: I'd stopped running just in time.

The memory – or, if it isn't a memory then the recurring image in my head – is associated not with fear exactly but with a sudden strong sense of the smallness of the individual body, the largeness of the world. Something destabilising but also invigorating. Like contemplating the stars arcing across the night sky, or the depth of the Mariana Trench, or the expanse of deep time – all the former worlds that hover just out of sight, momentarily screened by the everyday, the insistent rush of *now*, waiting to be brought back into the light.

ROCKS AND ICE

2

BOX 48903C16

On the evening of 2 December 2015, twelve icebergs, each weighing around 10 tonnes, were arranged in a circle in the middle of the Place du Panthéon in Paris's Latin Quarter. They had once been part of the Greenland ice sheet; now they were part of *Ice Watch*, a new art work by Danish/Icelandic artist Olafur Eliasson and the Greenlandic geologist Minik Thorleif Rosing, from the Natural History Museum of Denmark.

The Greenland ice sheet contains a unique archive of the past climate of planet Earth: the deep time climate. Eliasson and Rosing's art work was a chance to experience that archive at first hand, so the next morning – the first day the work was open to the public – I took the Eurostar to Paris. It was sunny and unseasonably warm in the city. The icebergs – several of them towering over Eliasson and his assistants – glistened in the sunlight, bringing to mind Neolithic standing stones or, perhaps, a giant clock face. Each iceberg had something of an individual personality. Some were milky, almost entirely opaque, and others clear enough to see the individual air bubbles trapped deep inside the ice. I sat on the steps of the towering neo-classical Panthéon to watch the scene. Behind me, inside France's great secular mausoleum, lay its dead, distinguished citizens: Voltaire, Victor Hugo, Marie Curie. In front of me the icebergs sweated. Viewed from a distance, they took on a cool, blue sheen – familiar and strange and beautiful all at once. Passers-by moved towards them. It was difficult to walk by without stopping to stare, to touch the ice, to run your hands across the uneven surfaces of the blocks.

Ice Watch was one of a series of art projects set up in public spaces around Paris by the organisation Artists 4 Paris Climate to coincide with the 2015 Paris Climate Change Conference (officially COP21 of the United Nations Framework Convention on Climate Change). Across the city from the Place du Panthéon the politicians and bureaucrats were at work on what would be known as the Paris Agreement. An international commitment 'to keep a global temperature rise this century well below 2 degrees Celsius above pre-industrial levels and to pursue efforts to limit the temperature increase even further to 1.5 degrees Celsius.'[1] Elsewhere in the city, Argentinean artist Tomás Saraceno hung two enormous silver and transparent plastic spheres from the ceiling of the Grand Palais exhibition centre. The work, titled *Aerocene,* envisages an emissions-free future where such creations might float around the world buoyed only by the heat of the sun and infra-red radiation from the Earth's surface. Across the Seine, in the Muséum National d'Histoire Naturelle, the Australian Janet Laurence filled a series of glass tanks with bleached coral and the skeletons of marine creatures, some wrapped in what look like white muslin shrouds, others attached to tubes or suspended in laboratory beakers. She called it *Deep Breathing – Resuscitation for the Reef.* (Eliasson and Rosing's icebergs, it should be noted, were fished out of the Nuup Kangerlua fjord in south-west Greenland; no glacier was harmed in the making of the art work.)

The last 2.58 million years of deep time have seen the planet cycle through a series of glacials (colloquially, ice ages) and interglacials. We are currently living in an interglacial, and the Greenland ice sheet is a holdover from the last ice age, when much of the Earth's surface was covered in a blanket of white, and mile-high glaciers extended over vast stretches of the northern hemisphere. In most places – Canada, Scotland, Scandinavia – the ice melted away just over 10,000 years ago, but in Greenland it persisted. The future of the Greenland ice sheet is now dependent on international co-operation and the outcome of summits such as COP21.

It's a familiar story: rising temperatures and glacial ice melting faster than it can be replaced by snowfall. It matters, we read in the papers, because if all the ice melts, global sea levels will rise high enough to submerge London and Manhattan; much, much more than enough to make the islands of the Maldives disappear. Melting Greenland ice is altering both the temperature and the salinity of the ocean, causing a weakening of the Gulf Stream. A significantly weakened system is likely to cause more severe storms in western Europe, raise sea levels on the east coast of the US and disrupt vital tropical rains.[2]

Culturally it also matters because if the Greenland ice sheet contains an archive of the past climate of planet Earth, when it goes, a lot of the Earth's past, our shared pre-human history, goes with it.

*

To learn more about this archive, I travelled to Copenhagen to meet Jørgen Peder Steffensen. A burly, shambling man, Steffensen is professor of glaciology and curator of the ice cores at the University of Copenhagen. On the day we met he had just come back from his twenty-ninth Greenland expedition and wore a thick beard and grubby white T-shirt. I mentioned studying Old Norse at university and Steffensen's eyes lit up. A new edition of *Snorri's Edda* – a thirteenth-century prose work about, among other things, the lives of the Norse gods – had just come out. 'It's the first new Danish translation since Victorian times, and it's much more juicy than the old one,' he said.

History has always been important to Steffensen. As a schoolboy he was good at maths and physics but really excelled in history classes: 'I got maximum marks, and everyone said you should study history and I said no, it leads to unemployment, but if you can qualify yourself in maths and physics then no one can take that away from you. It's like the new Latin.'

In Greenland, Steffensen was in charge of logistics for a team of eleven scientists and support staff. The purpose of the expedition was to relocate an entire research station, moving 150 tonnes of equipment, including a four-storey dome loaded on to skis, 465 km across the vast Greenland ice sheet. At the new site they laid out the future camp, placed the main dome in position, constructed garages and built a ski-way to allow skied aeroplanes to land. Photographs show the dome – a black, geodesic structure with a windowed turret on the top – like a dark diving bell floating on the white snow. 'In this line of work, even small incidents may have large consequences,' one of Steffensen's field plans reads. 'Even though we are scientists, we also share something with [sailors] – we are superstitious. Therefore we hesitate to mention specific incidents [that may go wrong] as this could become self-fulfilling.'

En route they collected snow samples, but it's the ice that holds the key to the scientists' work. 'Snow falls in the interior and it never melts away,' Steffensen explained. 'It just piles up into layers, and these layers are compressed into ice, and each ice layer contains information on the weather conditions of that year when the snow fell.'

Scientists collect this compacted snow in the form of cylindrical samples known as ice cores – the University of Copenhagen houses approximately 25 km of ice core, including the world's largest collection of deep ice cores (ice taken from more than 2 km beneath the surface). Drilling down through the ice layers, they travel back into deep time, and the ice they extract contains not just ancient water but even prehistoric air trapped inside bubbles deep within the ice sheet. At the very bottom of the Greenland ice sheet they are sampling the snow that fell more than 500,000 years ago. An ice core is a rare example of the possibility of direct contact with a past atmosphere. There's something a little fantastical, I think, about the way that these ephemeral-seeming things – air, water – have been preserved over such immense time-scales. If the ice core were melted, you could drink the water, which had fallen as snow 500,000 years ago, and breathe 500,000-year-old air newly released from a bubble in the ice.

The fundamental idea of using ice cores to study past climate was conceived by Willi Dansgaard in the early 1950s, in the same building in the University of Copenhagen where Steffensen's office is now located. Dansgaard, who died in 2011, was a palaeoclimatologist and an expert in precipitation. Presented with a sample of rainwater, he could study its isotopic composition and determine the temperature at which it had formed. (For example, a preponderance of heavier oxygen molecules indicates colder atmospheric temperatures.) Realising that the same method could be applied to snow and ice, he essentially invented the discipline of ice core climate science. During the 1950s and early '60s Dansgaard had to make do with samples taken from icebergs. Then, in 1964, he visited the US Camp Century military base in north-west Greenland to sample deep layers of snow, and discovered that the US Army Cold Regions

Research and Engineering Laboratory was drilling through the Greenland ice cap. As soon as the drilling was completed, he applied for permission to perform oxygen isotope measurements on the Camp Century core. The results of this study led to the first purely scientifically motivated deep ice-drilling project. Pursued during the 1970s and '80s by scientists from the US, Switzerland and Denmark, it was known as the DYE-3 drilling.

Steffensen still remembers precisely the moment his own career in glaciology began. Dansgaard was looking for manpower for DYE-3 and someone suggested he try the hardworking young physics student down the hallway. 'It was two in the afternoon on Friday, 4 July 1980, when Willi Dansgaard came and asked me if I wanted to go to Greenland and work on the drilling project for eight weeks. I thought it sounded exciting and was already on my way by Tuesday morning,' Steffensen said. On that trip he fell in love with glaciology, and with fellow scientist Dorthe Dahl-Jensen, the project chief of Steffensen's recent expedition. Today Steffensen and Dahl-Jensen are married and Dahl-Jensen leads Dansgaard's old research group. When their children were younger, they had to take turns, going out to Greenland on alternative years, but now they can travel together. Dahl-Jensen gets the money and Steffensen organises the expedition logistics. 'She writes very good grant applications, and I'm very good at burning through the money she throws my way,' he said.

*

The Greenland Ice Core Archive is stored in a dimly lit basement room kept at −28°C. Rows of cardboard boxes labelled with serial numbers and 'KEEP FROZEN' written on the sides were stacked on shelves running from floor to ceiling. There were trolleys for moving the boxes. The effect was something like the collection depot at Ikea.

Scientists studying the cores work backwards in time, counting the visible annual layers of compacted snowfall much as you might count tree rings. Combined with other methods such as geochemical analysis, this creates an ice-core-based time-scale. Through studying the chemical composition of the ice itself, they construct detailed pictures of the past climate of our planet, including information about local temperature, CO_2 concentration and, based on dust trapped in the ice, global wind patterns. The hope is that these data can help us understand how the climate system worked in the past, how it works today and how it might change in the future.

On Steffensen's first Greenland expedition in 1980 there was a fellow history enthusiast, a researcher called Henrik Clausen. 'He would say that we had now reached this layer and it corresponded to the time in which the French Revolution broke out and that they had been cold years, which meant poor harvests and hunger, or now we were in the years when the Vikings went out and settled in Iceland and Greenland,' Steffensen said. 'The ice cores could tell us about the climate in the different periods and what impact it might have had on the course of history. I was wildly fascinated.'

Historically minded visitors to the archive often ask to see a specific piece of ice. Shortly before my visit the US Ambassador had asked to see what Steffensen calls 'the nativity ice': the ice that fell as snow between 1 BC and 1 AD. 'So my colleagues and I, when we ran out of money one day we were thinking that we could probably melt that ice down and sell it in St Peter's Church for special occasions ...'

The ice that Steffensen was going to show me was far older

than the nativity ice. It came from a core known as NorthGRIP, where scientists have been able to count layers back as far as 60,000 years. (Deeper in the core, further in the past, the layering is less well defined and so not suited for visual counting methods.) He left me by the door and headed off down an aisle looking, he muttered, for box 48903C16. Watching him disappear, I tried not to think about all those films where people get trapped in frozen meat lockers: −28°C is very cold: the tiny hairs inside the nose begin to freeze, followed shortly by the ink in a biro. I stamped my feet as I waited. 'Our view as humans is a sort of compressing of the distant past,' Steffensen had said. In deep time we talk blithely of thousands, millions, even billions of years. Even in human time, in the history books, we skip happily across centuries of living – 'the Middle Ages', 'the Renaissance' – but the closer we get to the present day, the more time expands in our consciousness. If the human brain naturally compresses the past, then the scientists working with deep time are in the business of decompression. The recovery of time past. This was what I was waiting for.

Box 48903C16, when Steffensen reappeared and opened it, was lined with polystyrene. Inside, each 55 cm section of the ice core was wrapped in a numbered plastic bag. He lifted out bag 2712; we were looking at ice that had been buried at a depth of over 1,400 m – deeper than Ben Nevis is high; deeper than four Eiffel Towers standing on top of one another. It had been formed just over 11,700 years ago, during one of the very last winters of the Pleistocene Epoch.[3]

To view things egocentrically, the Pleistocene Epoch, which began 2.58 million years ago, is important because it is the geological unit of time during which *Homo sapiens* appeared, living as hunter-gatherers and co-existing with *Homo neanderthalensis*. When the ice we were looking at fell as snow in

Greenland, it would have landed on herds of mammoth and woolly rhinoceroses. Elsewhere, ancient Britain was still a peninsula attached to mainland Europe; North America was either buried under ice or a barren frozen wasteland of permafrost; small glaciers nestled in the mountains of southern Australia. Inside bag 2712 was the last ice core from the Pleistocene Epoch. The crystals glistened in the dim lighting. Running through the core were a series of distinct cloudy bands that Steffensen said were trapped particles of dust from Asia. He picked up a section of core from later in the sequence and held it out. Formed during warmer, calmer climatic conditions, the ice in this core looked quite different, much clearer, with just a few air bubbles. Between these two ice cores, the world changed for ever. The climate warmed rapidly, and the ice sheets retreated. Mammoths and other large mammals became extinct. So did the Neanderthals. The Holocene had arrived.

The Holocene (meaning 'entirely recent') began 11,700 years ago (with a counting error of ninety-nine years, and where the present is taken as the year 2000). The official transition is marked in the NorthGRIP ice core in a layer 1491.4 m deep. It is, at the time of writing, our own geological epoch, the part of deep time that we inhabit. The Holocene ice that I was looking at contained some of the first snowfalls of a whole new world. In the Holocene the nomadic hunter-gatherers would begin to develop farming and learn to live in settled societies. As a result, the population would increase and our ancestors would develop metallurgy, writing, money, the spinning jenny and, eventually, the fossil-fuel-burning internal combustion engine. Considered one way, we were looking at the beginning of civilisation.

*

In the 1970s Daansgard's ice core work turned up something completely unexpected. Back in his office, Steffensen opened a new chart on his computer. It showed that during the middle of the last ice age the temperatures in Greenland shot up by roughly 16°C during a period of less than twenty years.[4] It was as though Manchester had suddenly become Rio de Janeiro. Even stranger, the temperature suddenly yo-yoed down again, almost as abruptly. This happened not just once but many times. When Daansgard published his data, the scientific community was sceptical. If this was true, why hadn't it been noticed before? Where was the corroborating evidence? Was this just a technical glitch?

'There was no evidence simply because people were not looking. No one thought you should go into that much detail,' Steffensen said. Time had not been sufficiently decompressed, expanded to include enough detail to show such rapid fluctuations. Then in 1981 cores from the DYE-3 site confirmed Dansgaard's original findings and, over the next forty years, other climate records, including pollen deposits, ocean sediments and stalactites, revealed the same pattern. The temperature swings became known as Dansgaard-Oeschger events, after Dansgaard and a Swiss colleague, Hans Oeschger.

Sixteen degrees Celsius in less than twenty years. 'You cannot have that degree of change without having a regional and hemispheric climate reorganisation,' Steffensen said. 'And that means in Europe at the same time there must have been a change in wind patterns and rainfall to such an extent that it would have forced the animals and humans to move.' In less than twenty years the places that were good to live were no longer the places that were good to live. We tend to think of geological processes as slow, lumbering beasts, but here the end, when it came, was terrifyingly abrupt.

We understand that ice ages are triggered by small but complex variations in the Earth's orbit that alter the amount of sunlight hitting different parts of the globe. Dansgaard-Oeschger events, however, are weird. What causes them is still the subject of debate, though some scientists have suggested that rising levels of CO_2 could have reached a tipping point during the last ice age, triggering a series of chain events that caused temperatures to rise abruptly.[5] 'The climate system at certain times has been far from stable. So unstable that a tiny, tiny push would upset the whole load,' Steffensen told me. Though this is not evidence that current climate change will be as rapid, it is important because it shows that such a rapid change is possible. In comparison to the previous Pleistocene Epoch, our Holocene has been, in climatic terms, exceptionally stable. It was this stable climate that allowed farming to flourish and human society to expand and develop. This stability, we are learning, is not necessarily normal.

At the end of my visit Steffensen had to collect his daughter from school. We walked out together to the car park. The post-work traffic was picking up. Sunlight glinted off a windowpane, and an aircraft drifted overhead trailing a cloud of exhaust fuel. I asked Steffensen about the future. From the perspective of a palaeoclimatologist, what challenges might lie ahead?

'The farmer killed the hunter. We are all Cain's children now, and so we are much more sensitive to climate change than at any point during the human history,' Steffensen said. 'My greatest fear is that we inadvertently trigger a sequence of events that will affect food production, because all of a sudden it will stop raining where we need it to rain, and start to rain where we don't need it to … That's my biggest fear: that the change will be so sudden that farmers can't adapt.'

*

By 2 p.m. in the Place du Panthéon, the icebergs were sweating. Eliasson was much in demand and seemed harassed. I talked to a young man whose company makes jewellery in the shape of glaciers. 'We take a mould directly from the surface of the ice,' he said enthusiastically. A woman from Julie's Bicycle, a charity that helps arts organisations measure, manage and reduce environmental impacts, told me that the carbon footprint for *Ice Watch* was 30 tonnes of CO_2 – the equivalent of thirty people flying return from Paris, France, to Nuuk, Greenland. A child, out of sight of his parents, licked one of the icebergs.

Ice Watch, like many of Eliasson's works, first demands an almost physical, sensory response before allowing us to step back and think intellectually about the experience. What Eliasson calls the 'Wow' followed by the 'Aha'. And at COP21 one of the things he wanted people to think about was, of course, climate change. According to his web site:

> As an artist I hope my works touch people, which in turn can make something that may have previously seemed quite abstract into reality. Art has the ability to change our perceptions and perspectives on the world and Ice Watch makes the climate challenges we are facing tangible. I hope it will inspire shared commitment to taking climate action.

I watched as water from the icebergs dripped on to the paving stones and ran across the square. It was the sun. It was the hands of the passers-by warming the ice. Climate change writ small; climate change made tangible. Rosing, the Greenlandic geologist, towered over everyone. His father, Jens Rosing, had designed Greenland's coat of arms, and his own recent scientific work on the vastly ancient Greenland rocks led to the dating of the origin of life on Earth to several hundred million years earlier than previously thought. 'Scientists study the layers of the ice to learn about the past climate, but you can also see things happening in human society,' he told me. The Industrial Revolution shows up in the ice as a spike in levels of carbon dioxide, sulphur and methane. The invention of money (or at least the widespread adoption of coinage by the Greeks around the sixth century BC) is a spike in lead, a by-product of silver production. A dip in carbon dioxide levels in 1929 is the economic depression. The ice is an archive of human history during the Holocene. 'On the one hand it's telling us about a process [climate change] that is happening right now, but it also gives us the historical background to understand where this present has come from,' he said.

And in the deep future, too, our traces will be there in the ice. Just as we can look back now and find evidence of what was happening 20,000 years ago in the Pleistocene, so, if the ice in Greenland is preserved, any geologist who happens to be around 20,000 years hence will be able to look at it and judge the success of the Paris Agreement. Of course, if the ice has vanished then, that may, in itself, be an answer.

*

The Paris Agreement came into force on 4 November 2016 and immediately disappointed many people. There were complaints that the Agreement is fundamentally weak, since each country's contribution is essentially voluntary, and that, even if targets are met, we are still on track for dangerous levels of warming. The more hopeful might point out that evidence that international pacts can be effective can be found in the

Greenland ice sheet itself – a dip in sulphuric acid in 1985 is a direct consequence of the Rio protocol on emissions – and that after more than two decades of failed efforts to reach a global consensus on climate change, any agreement must still count as an historic achievement. Only Syria and Nicaragua failed to sign up: Syria was crippled by war at the time of negotiations, and Nicaragua argued that the deal was too weak.

Then in June 2017 Donald Trump pulled the US from the agreement, sparking consternation abroad and at home. 'If Trump Dumps the Climate Accord, the US is the Loser,' ran the headline on Bloomberg Businessweek. Even Exxon Mobil, the American oil and gas multinational, thought it a bad idea, and the states of California, New York and Washington all agreed to honour the decisions made in Paris. (At the time of writing it is expected that when Joe Biden, US President-elect, takes office in 2021, he will rejoin the Paris Agreement.)

I thought about what Steffensen had said when I asked him about climate-change denial. 'I'm constantly mesmerised by the fact that people are genetically put together in such a way that they will choose to disregard proof in order to have a simpler view of the world,' he said. 'For some, this world view includes a belief that religion and climate change are incompatible. I'm a good Christian in many ways, but in my mind there is space for science and Christianity and they are absolutely nothing to do with each other. Science was never meant to prove or disprove the existence of God. You can look at the sky, and knowing there's galaxies and dark matter and shit up there doesn't take away the sense of beauty and wonder.'

Meanwhile in Greenland, the ice sheet continues to melt. Melt happens every summer at the edges of the ice sheet, even when the ice sheet as a whole grows, but *Science* magazine reported that in 2016 'the melting started early and spread inland fast. By April 12 per cent of the ice sheet's surface was melting; in an average year the melt doesn't reach 10 per cent until June'. The pace of change has shocked researchers. 'Things are happening a lot faster than we expected,' geophysicist Isabella Velicogna said.[6]

In 2018 new work based on satellite observations and analysis of ice cores and models, showed that losses from the Greenland ice sheet had reached their fastest rate in at least 350 years.[7] In 2020 scientist Andy Aschwanden, wrote in the journal *Nature* that we are 'increasingly certain' that the twenty-first century will see 'unprecedented rates of ice loss from Greenland, unless greenhouse-gas emissions are substantially reduced.'[8]

*

By 3 p.m. the water was running downhill in the direction of the Eiffel Tower. People were taking selfies and photographing each other against the ice. You had to get very close before you felt the coldness of the frozen Arctic water.

I thought about the ice cores back in the basement in Copenhagen. The more Greenland ice we lose, the more important those cores will become. It's possible that one day the ice that I had seen in the basement in Copenhagen will be all that is left of today's 1.7 million km^2 ice sheet.

Eliasson, who had finished filming a segment with CNN, came over to where Rosing and I were standing. 'We were thinking that they would last through the conference,' he said gesturing towards the icebergs and shaking his head.

Rosing nodded: 'And now they might even be gone in only three or four days.'

3

SHALLOW TIME

If geologists had a Mecca, it would surely be the place called Siccar Point – a small, rocky headland at the base of a cliff in the border county of Berwickshire on the east coast of Scotland, around forty miles east of Edinburgh. You can find its picture somewhere near the beginning of many geology textbooks, and a bronze casting of the site is housed in the American Museum of Natural History in New York. Students of geology also know it as 'Hutton's Unconformity', after an eighteenth-century Scottish farmer and naturalist who would lay the foundations for the emerging science of geology and revolutionise our understanding of time.

From the small village of Cockburnspath, the path to Siccar Point runs along the coast between the soft grey of the North Sea and bare fields of red earth that glow against the damp green grass. Where yellow gorse blooms, the air smells sweetly, incongruously, of coconut. I was dressed in a waterproof jacket against the overcast sky. There was an Ordnance Survey map jammed into the plastic map case, which Jonny refused to get involved with. My husband grew up in the countryside and despises brightly coloured rainwear and map cases and official walking gear. I was a Girl Guide and call these things being prepared.

The path led past the rubbly ruins of St Helen's Chapel, silhouetted against the white sky, and past the entrance to a vegetable-processing plant before climbing sharply uphill to the top of a steep, almost sheer, grassy cliff that runs down to the bare rocks of Siccar Point far below. James Hutton, alongside his friends John Playfair and Sir James Hall, visited this spot by boat in 1788. From the landward side, I realised, we would need to scramble down the steep green cliff, the final section of the pilgrim's trail a long, thin slick of red mud. 'Safety warning,' read a sign. 'The slope down to the rocks is steep and dangerous. Please proceed with care at your own risk.' This did not inspire confidence. Someone had tied a red rope to a rickety wooden fence post; it would be possible to cling to it as you inched downwards, trying not to look at the sea-swept rocks.

'Well,' one of us said at last. 'Shall we go?'

*

Time used to be shallow. 'The poor world is almost six thousand years old,' Rosalind tells Orlando in Shakespeare's *As You Like It*.[1] In the year 1623, when the First Folio was published, this was the current orthodoxy, upheld both by playwrights and by great scientists such as the German astronomer Johannes Kepler. In 1658 James Ussher, archbishop of Armagh and primate of all Ireland, settled the matter definitively in *The Annals of the World*.[2] Using the chronology of the Bible and tying that to dates in world history, he was able to announce that the world began on the 22 October 4004 BC. He even specified the day of the week – a Saturday – and a time – around six o'clock in the evening. (In this project, as the historian Martin Rudwick points out, Ussher was not what we would term today a 'creationist' or 'Young Earther' but a 'public intellectual in the mainstream of the cultural life of his times' – someone using the most current scientific theories of the day.[3] Isaac Newton was among the other thinkers who had attempted this feat.) Some fifty years after Ussher's death his chronology was included in the margins of the new King James translation of the Bible, gaining canonical weight that would endure for 200 years.

But by the middle of the eighteenth century the cracks in the story were beginning to show. 'The Chinese, for example, mocked the story of Noah's flood, which was supposed to have occurred around 2300 BC,' the biographer Stephen Baxter has written.[4] European missionaries discovered that 'Chinese written history stretched back for centuries *before* this date and made no mention of a disastrous global deluge.' In Europe scientifically minded thinkers began asking awkward questions of their own. In France the naturalist Georges-Louis Leclerc, Comte de Buffon, became convinced that the Earth was far older than 6,000 years. He set out to prove it.

The seventeenth-century German polymath Gottfried Wilhelm Leibniz had speculated that the Earth was originally a molten ball and retained a molten core. This fitted with what Buffon had himself observed – for example, that temperature increases as you descend into underground mines. If Buffon could work out the rate at which such a molten ball would cool, he could then work out the age of the Earth. In the 1760s he began a series of experiments heating iron balls to near melting point and measuring the time it took them to cool. Not trusting the crude thermometers of his day, he measured instead the time it took for the ball to cool sufficiently to be held by hand for a minute without damage: 'occasional injury was part and parcel of the process,' the stratigrapher and palaeontologist Jan Zalasiewicz has written. 'Worse, it seems that Buffon regarded women's hands (more sensitive, you see) as the best measuring devices.'[5] In 1778 Buffon published the results in his most famous book, *Des Époques de la Nature*, which Zalasiewicz, who is translating Buffon in his spare time, points to as the first science-based narrative of the Earth. In it Buffon announced the new age of the Earth: not 6,000 but a shocking 75,000 years old.

Writing at a time when challenging prevailing religious orthodoxies could be career suicide, the pragmatic Buffon's revolutionary book included a section arguing that his 'purely hypothetical ideas' concerning the Earth could in no way harm the 'unchanging axioms' of religious faith, which were 'independent of all hypothesis'. On the whole, Zalasiewicz writes, 'the stratagem worked' and Buffon avoided trouble and scandal while privately believing the true age of the Earth to be even older than 75,000 years: some of his unpublished manuscripts suggest that 3 million might be closer.[6]

Even at 3 million years, Buffon's Earth was far too young, but his was an important early attempt to envisage an Earth that existed within deep time – a concept without which our modern theories of geology, physics, astrophysics and even evolution would be impossible. Darwin's theory only makes sense given enough time for evolutionary changes to take place.

In 1778 time was still too shallow. It was this challenge that James Hutton would take up ten years later at Siccar Point.

*

The Edinburgh guidebook I'd borrowed from the library had left James Hutton's name off a list of Edinburgh's famous personages. On the Royal Mile there is no grand statue to match those of his Enlightenment contemporaries the economist Adam Smith and the philosopher David Hume – a rather florid rendering of the latter clad in a Roman toga, one flabby breast displayed along with a toe, burnished gold from where philosophy students rub it for good luck before exams. He is absent from a list of noteworthy burials on the web site of the Greyfriars Kirk in the Old Town, where he was interred. On a damp and overcast morning we searched for the grave. The kirkyard was filled with Spanish tourists taking selfies in front of the polished pink granite headstone of a Skye terrier known as Greyfriars Bobby who, according to popular legend, spent fourteen years guarding the grave of his dead master. The terrier was clearly a bigger draw than James Hutton, who was nowhere to be seen at all. Wandering inside the kirk, we fell into conversation with a white-haired churchwarden clearing up after the Sunday service. He was pleased we'd come to look for Hutton – he himself was a former physics student, and proud of the connection between his kirk and the great man. He told us to wait while he fetched the keys.

Hutton's grave lies out of sight in a quiet, gated and walled section of the kirkyard composed of two rows of individual family plots. He is in the Balfour plot by virtue of his mother. For many years the grave contained no reference to Hutton, but in 1947 a simple pale grey granite rectangle was attached to the red brick wall: 'James Hutton H.D. F.R.S.E., 1726–1797, The Founder of Modern Geology'.

It seemed sad, I told the churchwarden, that for so long one of the most important thinkers of the eighteenth century lay in an unmarked grave, and that even now he had such a modest memorial. The warden tilted his head to one side. 'It depends on whether you think death is just the end or the route to somewhere else,' he said. Or, as Hutton himself put it: 'We are to consider death only as a passage from one condition of thought to another.'[7]

James Hutton was born not far from Greyfriars Kirk, in 1726. His father, a merchant and Edinburgh City treasurer, died when he was three, leaving two farms in Berwickshire and certain responsibilities to provide for the family. After studying medicine in Edinburgh, he entered into a relationship with a Miss Edington, and in 1747 she gave birth of his only child, a son. Not much is known about this part of Hutton's life or the fate of Miss Edington – apparently Hutton had little to do with his child beyond providing financial assistance. He continued his medical studies in Paris and Leiden, but rather than practice medicine spent 1750 to 1754 travelling and studying agricultural methods in East Anglia and then the Low Countries. In 1754 he moved to his lowland farm Slighhouses, eager to implement the modern agricultural practices he had learned. Some biographers have speculated that the scandal of his illegitimate child, and the pecuniary pressures he now faced as head of his family, were contributing factors in his decision to leave Edinburgh. Whatever the truth, the move would lead him to his life's work.

Accustomed to the intellectual life of the city, at Slighhouses Hutton was often miserable. Determined to make a success of the farm, he engaged in hours of back-breaking physical labour, hauling stones from the fields and digging drainage ditches. During his travels in East Anglia and the Low Countries he had begun increasingly to take note of the rocks and landforms around him. Working at Slighhouses, he turned that same attention to the Scottish lowlands. 'One of the difficulties Hutton faced was a lot of soil erosion,' Colin Campbell, chief executive of the research centre the James Hutton Institute, has said. 'He was forever wondering how to keep the soil on the land and stop it disappearing off in rainstorms down the rivers.'[8]

Scientific orthodoxy at the time provided no explanation for the continued building up of the Earth – only its destruction. Creation, as the Bible taught, happened only once. If this were true, all mountains and then, eventually, all land, would wear away. But Hutton began to believe that there *was* a renewal process. What if soils and eroded rock fragments, such as those he struggled against on his farm, came together again? What if they eventually consolidated to form new rocks? What if land was being created as well as destroyed?

*

Between 280 and 220 million years ago the Moray Firth basin in Scotland and much of central England was a hot, dry desert. Across the globe, sea levels had fallen, the water retreating to reveal vast, sandy plains. Sand dunes formed alongside immense salt flats similar to those found in the Persian Gulf today. Some of that sand became the rock we call New Red Sandstone – though the colour is more a warm, muted rose. In the late 1800s it was chosen to build the new Scottish National Portrait Gallery, a turreted Gothic Revival building on Queen Street in Edinburgh. A small statue of Hutton can be seen on one exterior wall, but most people passing by, the Hutton scholar Alan McKirdy told me, are unaware of who the statue represents. Inside there is a portrait of Hutton painted by Sir Henry Raeburn. One afternoon I went there to look for it.

Descriptions of Hutton's character range from ascetic to bawdy. His younger friend and first biographer, John Playfair, tells us that he 'ate sparingly, and drank no wine', had a 'thin countenance' that 'bespoke extraordinary acuteness and vigour of mind' and had 'that genuine simplicity, originating in the absence of all selfishness and vanity, by which a man loses sight of himself altogether, and neither conceals what is, nor affects what is not'.[9] Stephen Baxter, in his book on Hutton, argues instead that Hutton's letters show us someone 'warm, impulsive, crude, funny, lustful and frequently drunk'.[10] McKirdy writes that 'he liked brandy toddy'.[11]

Raeburn's portrait shows him dressed in a brown jacket, waistcoat and breeches, wigless, with a long nose, domed forehead and receding hairline. A table holds several crudely painted rocks and fossil shells, alongside an unpublished manuscript entitled 'Elements of Agriculture' – Hutton's collective wisdom on his fourteen years as a farmer. Robert Louis Stevenson wrote of the portrait: 'Hutton the geologist, in Quakerish raiment, looking altogether trim and narrow as if he cared more about fossils than young ladies.'[12] A nearby portrait of his friend Joseph Black shows the chemist with light radiating from behind his bewigged head, facing the viewer, holding a glass tube aloft, arms spread; Hutton sits awkwardly sideways on a chair, legs crossed, hands folded neatly in his lap, gazing mournfully away from us. Black looks as though he's been caught mid-lecture; Hutton looks like he'd rather be somewhere else.

Around the time this portrait was painted, the forty-one-year-old Hutton returned to Edinburgh to live with his sisters. Any scandal associated with his name had presumably died down, and his financial prospects had improved. Years earlier he and an old friend, James Davie, had devised a way to transform the soot collected by Edinburgh chimney sweeps into sal ammoniac, the salt of ammonia and hydrogen chloride, used in Hutton's time in the dyeing industry and for working with brass and tin. With the salt previously available only as an import from Egypt, Hutton and Davie's business eventually turned a handsome profit. Finally freed from the need to earn his living as a farmer, he found himself in a city at the heart of the intellectual ferment of the Scottish Enlightenment. A list of some of his friends and contemporaries reads like a Who's Who of the great and the good: Hume, Playfair, Smith, Black, the engineer James Watt and the poet and lyricist Robert Burns. With Smith he founded a gentlemen's supper club called the Oyster Club. Claret was the drink of choice, Baxter writes, and 'a gentleman would be labelled a two- or three-bottle man, depending on his consumption over dinner.'[13]

In a study so full of fossils and chemical apparatus of various kinds that there was barely room to sit down, Hutton continued to brood on the theories he had begun to develop at Slighhouses. What force, he wondered, could possibly transform eroded rubble and particles of sand into new rocks? How could these rocks, formed under water, have been raised up to create new land?

According to Playfair, the answer had come to him by the early 1780s. At first he would only discuss his surmises with Playfair, Black and another friend, John Clerk. 'He was in no haste to publish his theory; for he was one of those who are much more delighted with the contemplation of truth, than with the praise of having discovered it,' Playfair wrote.[14] But in 1785 his paper 'Theory of the Earth' was read at two meetings of the Royal Society of Edinburgh. He had found an answer to his dilemma: heat.

Hutton argued that the Earth was a perpetually renewing system engaged in an endless cycle of destruction and repair – one that could be explained by understanding physical processes operating in the present day. He proposed (like Buffon) that the centre of the Earth was a molten ball. It was heat from this molten centre that provided the mechanism to create new rocks. Some were melted and then cooled to form what we know as igneous rocks such as granites, while in the ocean sedimentary rocks were consolidated through a baking process. Today we know that it is also heat, driving plate tectonics, which is responsible for the movements of continental land masses and, as a result, the raising up of the land.

According to the geologist and broadcaster Iain Stewart, over 200 years ago 'Hutton got almost everything correct.' [15] In Edinburgh, however, his audience reacted with a mixture of indifference, misunderstanding and hostility.

Hutton claimed that the sedimentary origin of so many rocks was evidence of a past 'succession of worlds' – a system apparently designed to maintain a habitable Earth. It is to this that he refers in what is perhaps the most famous sentence in geology: 'The result, therefore, of our present enquiry is, that we find no vestige of a beginning, no prospect of an end.' [16]

Alan McKirdy and Donald B. McIntyre have written that 'His opponents distorted this statement, pretending that Hutton claimed [...] that there had been no beginning and would be no end.' [17] To his great distress, he was accused of atheism. But for Hutton, it was the notion of an Earth doomed to destruction that seemed heretical, incompatible with the idea of a loving God.

Stung by his experience at the Royal Society, the sixty-year-old Hutton realised that to convince people of his theory he would need to find more proof – and to gather this proof he would need to return to the rocks or 'God's books', as he called the specimens he collected, 'wrote upon by God's own finger'. [18]

*

At the bottom of the cliffs at Siccar Point I stood on the flat shelf of rocks that Hutton had visited back in 1788. The sky was white and heavy. The sound of the waves on the rocks was a constant shushing and the water was the colour of salt-encrusted green glass.

By this point in his search for evidence, Hutton had already visited several other locations in Scotland, including a three-day journey to Glen Tilt in Highland Perthshire, where he found veins of pink granite penetrating grey metamorphic rock, proving that the pink granite must have been molten when it came into contact with the grey rock, and that it must therefore be younger than the grey rock. (The age of granites was another aspect of Hutton's theory that had enraged his audience – conventional wisdom held that they were among the oldest rocks, not the youngest.) In the years that followed the Glen Tilt expedition, Hutton and Hall conducted a series of experiments to find further evidence of the role of heat and pressure in the formation of rocks. As they worked, Hutton became increasingly aware that if his theories of destruction and renewal were correct, then the amounts of time necessary to accommodate these processes must be incredibly, stupendously vast. It was to find further evidence to support this new claim that he set out by boat one day in June 1788 to explore the bare rocks and grassy cliffs near Cockburnspath, and came upon Siccar Point.

At the Point the rocks at first looked all the same to me – dull greyish brown, spotted here and there with mustard-yellow lichen. I had to wait a bit to get my eye in. Slowly, they began to reveal themselves. Of course it was my perception that was changing, but it felt as though it was the rocks themselves – like watching a Polaroid develop, or adjusting the brightness on a digital file. There were two rock types, distinct in colour and form. One was dull, dusty pink, the colour of fresh plaster. It lay in horizontal layers. Below it was a mottled dark and light grey rock with a faint bluish tinge. Here, the layers were nearly vertical – like a sheaf of papers in a hanging filing cabinet. When Hutton arrived by boat with Hall and Playfair, he realised that these two rock formations told a story of epic, violent proportions.

The grey rocks are a form of sandstone – greywacke – originally formed as horizontal layers on the sea floor, then wrenched upwards and tilted ninety-five degrees. After that, time passed. A lot of time. A new world with new rocks formed above the greywacke. These in turn eroded and settled on the grey rocks. Eventually this new material was in its turn consolidated, hardened, transformed into a new rock – the Old Red Sandstone. Hutton could observe these processes of erosion and sedimentation operating around him in 1788 – an observation originating from his time at Slighhouses – and the slowness with which they operated was proof that it would have taken far, far longer than 6,000 or even 75,000 years to create the landscape of Siccar Point.

Staring at the rocks, Hutton had a vision of the slow and endless cycling of the Earth. As Playfair later wrote: 'The mind seemed to grow giddy by looking so far into the abyss of time.'[19]

*

In 1791 Black wrote to Watt, telling him that Hutton was very ill and in great danger. Though he recovered, the affliction – probably a prostate problem – marked a decisive shift in Hutton's health. For the rest of his life he was frequently ill.

At that time he was still living in Edinburgh, in a house just off the Pleasance and just beyond the Old Town on St John's Hill. The building has long since disappeared, but in 1997 the site – a piece of waste ground since the 1960s – became the

Hutton Memorial Garden. It is a strange place: a small patch of gravel littered with cigarette butts and hedged by dark green rhododendrons, squashed between a university accommodation block and the back of a multi-storey car park, where ventilation ducts constantly rattle. Two boulders showing granitic veins come from Glen Tilt and illustrate Hutton's work on the origin of granite; three others are conglomerates (coarse grained sedimentary rocks) from Dunblane, representing Hutton's understanding of the cyclical nature of geological processes.

On that spot on Saturday, 26 March 1797, Baxter tells us, Hutton woke in a great deal of pain. He tried to work, making some notes about a new naming system for minerals, but the pain worsened. That evening medical attention was sought, but by then it was too late. Lying in bed, surrounded, perhaps, by his geological papers, Hutton's last action was to reach out his hand towards the doctor coming though the door.[20]

*

When James Hutton helped shift human thinking from a shallow- to a deep-time world, he was participating in something as fundamental, as profound, as the work of Nicolaus Copernicus, Galileo and Kepler, when they argued that the sun, not the Earth, is at the centre of our solar system. But Hutton's great theory was never widely read or understood in his lifetime. A dense, sprawling prose style is often blamed. Even Playfair, one of his biggest fans, talked despairingly of the 'prolixity and obscurity' of Hutton's writings.[21]

His work might perhaps have vanished altogether were it not for the actions of the men who had accompanied him to Siccar Point. After Hutton's death, Playfair – playing Boswell to

Hutton's Dr Johnson – wrote a flattering biography (1803) and a sort of nineteenth-century Very Short Introduction guide to his great theory: *Illustrations of the Huttonian Theory of the Earth* (1802). Meanwhile, in 1824, Sir James Hall returned to Siccar Point with an enthusiastic young man from Strathmore who was about to become the most eminent geologist in Britain in the first half of the nineteenth century – Charles Lyell.

Lyell, whose work was to form much of the basis of modern geology, developed and further popularised Hutton's thinking in his *Principles of Geology* (1830–33). Written in stirring, romantic prose, the *Principles* was a nineteenth-century blockbuster. A first print run of 4,500 copies quickly sold out, followed by a hasty reprinting. In contrast to Hutton, Lyell became something of a celebrity – the Brian Cox of his generation. When he spoke in Boston, Massachusetts, more than 4,000 people tried to get tickets.[22]

Charles Darwin wrote in *On the Origins of Species* that 'He who can read Sir Charles Lyell's grand work, which the future historian will recognise as having produced a revolution in natural science, yet does not admit how incomprehensibly vast have been the past periods of time, may at once close this volume.' [23] Through Lyell, Hutton's idea that processes currently operating on the Earth could explain Earth's history became known as Uniformitarianism, summarised for generations of geology students as 'the present is the key to the past'.

But Lyell also had his detractors, and many of them were supporters of a competing theory of Earth's history. Catastrophism was associated with the work of the great French palaeontologist Georges Cuvier, often said to have 'invented' extinction by showing that the mammoth is both different from the elephant and not alive on Earth today. Using Cuvier's work, catastrophists argued that the history of the Earth was not one of slow, steady cycles but of a series of sudden, rapid or cascading events, such as volcanic eruptions and mass extinctions.

In the end, both turned out to be somewhat right. Today most geologists combine the two theories and come up with a picture of the world of deep time as a series of slow steady revolutions periodically punctured by local, regional or global catastrophic events.

*

Hutton saw 'no vestige of a beginning, no prospect of an end', but in the nineteenth and twentieth centuries efforts were made to put an actual number on the age of the Earth. In 1897 the Scottish mathematical physicist Lord Kelvin, assuming, like Buffon, that the Earth commenced as a molten body and cooled at a steady rate to its present state, gave an estimate of 20–40 million years. In 1900 the Irish physicist and geologist John Joly measured the amount of salt in the ocean and upped Kelvin's estimate to 90 million years. Scientists were getting closer, but the number was still far, far too small.

It was the discovery of radioactivity around the beginning of the twentieth century that transformed the debate. In 1913 the English geologist Arthur Holmes produced the first geological time-scale using radiometric methods. Dating rocks based on the radioactive decay of uranium to lead, he calculated that the oldest rocks discovered on Earth were at last 1.6 billion years old. By the 1950s this same method was being used to suggest a date for the age of the Earth of 4.5 billion years, and current thinking puts it at 4.6 billion – a figure that would have been unthinkable to Buffon, Hutton, Lyell and their respective contemporaries.

The last time I visited the Natural History Museum in South Kensington, the curators, trying to convey this immense age to the general public – and humankind's place within it – had fallen back on the familiar clock-face analogy: if 4.6 billion years of deep time is represented by twenty-four hours, then humans won't appear on the scene until two minutes to midnight.

Such analogies tend to emphasise our temporal insignificance, and though Lyell died around sixty-five years before Holmes published his radiometric work, he was probably one of the first people to appreciate that, in the context of deep time, the lifespan of an individual species counts for little.[24] But for Lyell, finishing Book One of the *Principles*, this insignificance was only a further source of wonder. Celebrating the fledgling science of geology and the world of deep time it was uncovering, he wrote gleefully that, 'although we are mere sojourners on the surface of the planet, chained to a mere point in space, enduring but for a moment of time, the human mind is not only enabled to number worlds beyond the unassisted ken of mortal eye, but to trace the events of indefinite ages before the creation of our race'.[25]

*

At Siccar Point we crouched on the rocks beneath the high, grassy cliff and the flat, white sky. Here, a little over 230 years ago, shallow time had deepened. We bent down and placed our hands on what geologists call the contact – the place where two different layers of strata meet – in this case the greywacke and the Old Red Sandstone.

Underneath our palms around 30 million years of time was missing. Perhaps during this period, the transition between the Silurian and Devonian worlds, no rock was laid down. Or perhaps the rock that had been there had eroded away. In the ledger of deep time more is lost than is retained.

Then for scale we took photographs of each other next to the contact. Two small figures in a bright blue and a black raincoat. Individual human bodies attempting to make comprehensible the incomprehensibility of deep time.

4

THE AUCTIONEER

On the morning of 12 February 1947, the artist P. J. Medvedev was at work on a painting of the town of Iman (now Dalnerechensk) in the far east region of Russia. Moments later an immense fireball brighter than the sun streaked across the sky above the nearby Sikhote-Alin mountains, leaving behind a 33 km smoke trail. At 10.38 a.m. a huge explosion rent the air, seen and heard from as far as 300 km away.[1] Medvedev picked up his brush.

The resulting picture shows a glowing, pearlescent smoke trail narrowing to a bright fiery point over steep-roofed, snow-banked houses. He had just produced a painting of the largest meteorite fall in recorded history. The image was later made into a Soviet stamp to celebrate the tenth anniversary of the fall.

It took a team of Soviet geologists three days to reach the impact site, buried deep within a remote snowy forest of pine, spruce and larch. The meteor was an estimated 600 m in diameter. When it exploded in mid-air – the result of meeting air resistance as it entered the Earth's atmosphere – the blast flattened trees in every direction and left impact craters measuring up to 26 m across. Over 21 tonnes of extraterrestrial material fell on the surrounding countryside.

Had the meteor exploded over Moscow or London instead, this story would be very different. Thanks to the sparse population of Russia's far east, there were no reported casualties. A curiosity, not a calamity.

Sixty-two years later at Christie's auction house on King Street in St James's, London, I watched a small crowd bidding for meteorites. In fact, most of the people on the floor were sellers come to say a last goodbye to their treasures – but the phones were busy, along with the internet bidders live-streaming the event from around the world.

On that morning you could buy, among other things, a slice of the planet Mars (technically a Martian meteorite, a lump of rock knocked off the planet's surface during an asteroid strike), a petrified tree stump from Oregon, a *Tyrannosaurus rex* tooth from Montana and a lump of acid-yellow crystalline sulphur from Bolivia.

A tanned, grey-haired man in a dark blue polo shirt kept raising his paddle. He bought five meteorites, a *Triceratops* horn and a *Tyrannosaurus rex* toe claw, spending a total of £44,000, not including the buyer's premium (25 per cent of the hammer price up to £225,000). His most expensive single purchase – £14,000 – was Lot 30: a 22 × 28 × 10 cm, pewter-hued, shrapnel-like, Sikhote-Alin meteorite. I guessed that he worked for a museum and was putting together a collection but no, he told me, when we spoke after the auction, the meteorites were just a personal interest. A fancy, really. 'For me it's an emotional thing. It's so strange that they come from a long way away and travel all the way here.' I couldn't place his accent. Italian? Greek? He was holding a small black bag, like a cross between a briefcase and a handbag. 'The dinosaurs I like because they remind me of my kids when they were young.'

And what would he do with the meteorites? They were

destined for display on the wall of his home office, 'just like a painting'. He wasn't an expert, he said, moving towards the door of the auction house. Far from it. In fact, he gave an embarrassed laugh, two of the meteorites he'd bid for by mistake.

※

According to many geologists, the oldest Earth rock is the salmon-pink, ebony-banded Acasta Gneiss from Canada's Northwest Territories. At around 4 billion years old, it is just half a billion years younger than Earth itself. (A sample was sold at the Christie's auction for £900.) Slightly older, at around 4.4 billion years – and not technically rocks – are the zircon crystals found embedded in the Jack Hills in Western Australia. To go beyond this, to go further back into deep time, you have to go extraterrestrial.

A week after the sale, I returned to Christie's to view another Sikhote-Alin fragment, this one the colour of pencil lead, dark and gleaming and covered with tiny, rounded indentations, as though it had once been a ball of soft clay pressed all over by a small, insistent thumb. There, and there, and there. Made almost entirely of metal, it is thought to be the core of a melted asteroid formed during the birth of our solar system. When I held it in my hand, I was touching the oldest thing I will ever encounter: a third of the age of the universe itself. For more than 4.5 billion years, while Earth was cycling through ice ages and greenhouse worlds, this lump of rock drifted in darkness, through the cold spaces between the planets and the asteroids. Now it was here in a busy auction house in St James's, labelled and given a reserve price of £15,000 to £25,000.

'I think about how old it is and it just blows my mind,'

James Hyslop, Christie's head of Science and Natural History, said to me. In 2016 Hyslop developed Christie's first dedicated Science and Natural History sale – now a regular event. On the day we met, the auction house was gearing up for 'handbags and accessories'. The rocks, dinosaur bones and ammonites from Hyslop's pre-sale exhibition had been replaced with Gucci, Hermès and Louis Vuitton. One value set replaced by another – though putting a price on things at auction brings a strange sort of equivalence. For the cost of the Sikhote-Alin meteorite fragment you could buy, at other Christie's sales that year, a Hermès magnolia Togo leather Birkin 25 with palladium hardware handbag from 2018, a letter written by Leonard Cohen to Marianne Ihlen in 1961 or a 1926 drawing by Paul Klee titled *Attacking Plants*.

'In my opinion meteorites are extraordinarily undervalued,' Hyslop said. At auction they have yet to break the million-dollar threshold, though some have sold for that amount privately. 'And when you think that the total known mass of meteorites in the world is less than Earth's annual gold output, as a rare substance, they should be higher up there. I find that astonishing.'

Because a few meteorites do sell for six figures, the average price at auction is around £11,000, though at the sale I attended there were plenty going for between £1,500 and £5,000. Part of the reason for these low prices, Hyslop speculated, is that the art market is largely unaware such things are even for sale. Most of the meteorites that come to auction are 'primary', meaning that they've not been auctioned before, having either been sold privately or coming directly from the field. 'So you can buy truly world-class meteorites quite readily at the moment.'

What is it that makes a world-class meteorite? The diamond industry has the four Cs: cut, colour, clarity and carat.

Auctioneers of meteorites, it transpires, have the four Ss: size, shape, science and story.

When it comes to size, bigger is better – up to a point. 'You get to a certain size when it just becomes too difficult to move around, and the price stops going up until you reach a really, really, wow impressive size and it goes up again,' Hyslop said. A meteorite the size of the table we were sitting at would have less commercial appeal than, say, something a third that size. 'But a meteorite the size of this room' – we were sitting in a space around the size of a large changing cubicle – 'would be worth a fortune.'

Shape is probably the biggest driver of value. 'Some iron meteorites look just like contemporary abstract sculptures, but if you have one that's a little bit bricky and blobby it's not going to be so easy to sell,' Hyslop explained. Aesthetics play a big part in the meteorite market; buyers want something that will look attractive around the house, or that corresponds to their idea (probably from a movie) of what a meteorite *should* look like.

Then there's the science. A rock falling from the sky cannot be officially named as a meteorite until it's been recognised by the Meteoritical Society. Then a report must be published in their journal, *The Meteoritical Bulletin*, and a sample deposited at an internationally recognised institution, such as the Natural History Museum in London. When it comes to auction, Hyslop wants to know if the meteorite is of particular scientific note, how rare it is and where it has come from. (Lunar and Martian meteorites are particularly sought after and particularly rare. Randy L. Korotev, a meteorite expert from Washington University in St Louis, estimates that fewer than one in a thousand of all known meteorites are of lunar origin, and that the number is about the same for Mars.[2])

Finally, Hyslop will ask if there's a story attached to the meteorite. Meteorites from historic falls – such as the Sikhote-Alin meteor shower – do well at auction. An otherwise undistinguished meteorite is prized because it fell near the waiting Russian artillery forty-eight hours before the Battle of Borodino of 1812 – during the bloodiest of the Napoleonic Wars. 'Another example: there are only a handful of meteorites that have landed in the UK. If you're a British collector, you'll pay a premium for that story of it landing in a field in Yorkshire,' Hyslop said. 'Or there's one meteorite that killed a cow in Argentina. It's the only confirmed kill from a meteorite, so again there is a premium on that particular specimen.' Faced with the majestic expanse of deep time you can't help noticing that we often show ourselves to be focused primarily on the attractions of novelty and trivia above all else.

Hyslop held the meteorite up to the light. 'For the art market meteorites are a big *memento mori*. We're pretty certain it was a meteorite that did for the dinosaurs, and Hollywood loves reminding us that a meteorite could do for us.' At his first meteorite sale back in 2014, one of the biggest client list crossovers was with the collectors of *memento mori* and *vanitas* paintings – those meticulously detailed still-lifes of full-blown flowers, glistening piles of grapes, snuffed out candles, silent musical instruments and ivory-coloured skulls, symbolising the inevitability of our demise and the transience of earthly pleasures and achievements. 'Ultimately meteorites act as a symbol of death,' he said.

Later, walking along King Street, I found myself glancing upwards at the sky. In St James's Square I took out my phone. On eBay there were meteorites for sale. A practitioner of English witchcraft and magic was selling 2 cm fragments of the Sikhote-Alin meteorite for £15 a go. For £4,420 a Floridian member of the International Meteorite Collectors Association could sell you a 1.16 kg pallasite meteorite from Kenya. Pallasite meteorites contain extraterrestrial gemstones – in this instance, olivine crystals like drops of honey set in a silvery metal matrix. If I owned a pallasite meteorite, I'd want to extract one of the stones to hang on a pendant or set in a ring.

From there, following the looping logic of the internet, I began reading someone's theory that Stonehenge was a prehistoric meteorite predictor,[3] and then about its twenty-first-century equivalent: the ATLAS asteroid impact early warning system developed by the University of Hawaii and NASA.[4] I read that the estimates for the mass of material that falls from space on to the Earth each year range from 33,600 to 70,800 tonnes;[5] that most of the mass comes from dust-sized particles; that every 2,000 years or so a meteoroid the size of an American football pitch hits Earth and causes significant damage;[6] that asteroids with a diameter of 20 m strike Earth approximately twice every century;[7] that the Ancient Egyptians made jewellery out of iron-rich meteorites;[8] that in 1954 a housewife Ann Hodges was napping on her couch in Sylacauga, Alabama, when a cricket-ball-sized hunk of black rock broke through the ceiling of her rented house, bounced off a radio and smashed into her hip. A photograph from the time shows Ann lying in bed while a doctor pulls back her night dress to reveal a large black bruise, about the size and shape of a rugby ball and feathered at the edges like ink running on damp paper.

Ann became famous for a while. There was much back and forth as to who owned the meteorite. The Hodgeses' landlady claimed it as hers; Ann felt that 'God intended it for me'.[9] Eventually the landlady settled out of court in exchange for $500. The Hodgeses had hoped to make money from the meteorite, but by the time they'd reached the settlement, interest had waned and no purchaser was forthcoming. Eventually they donated the specimen to the Alabama Museum of Natural History. Ann later suffered a nervous breakdown and the Hodges divorced.

In 1972, at the age of fifty-four, Ann died of kidney failure. Speaking to the press, her ex-husband, Eugene, blamed the meteorite for Ann's breakdown and the collapse of their marriage. She 'wasn't a person who sought out the limelight,' the museum director Randy McCredy told a *National Geographic* reporter. 'The Hodgeses were just simple country people, and I really think that all the attention was her downfall.' [10]

5

TIME LORDS

Like almost everyone alive today, Philip Gibbard – Emeritus Professor of Quaternary Palaeoenvironments at the University of Cambridge and secretary-general of the International Commission on Stratigraphy (ICS) – was born in the unit of geological time we call the Holocene Epoch. But in 2018, the year before he turned seventy, the ICS announced the first major change to 'our' portion of deep time since 1885, when the Holocene was officially recognised at the International Geological Congress in Berlin.[1] The normally staid world of stratigraphy – the branch of geology concerned with the order of rocks and their place in the geological time-scale – became a storm of expletive-spouting professors, distressed climatologists and enraged geographers sounding off to the press. This was very much not business as usual.

I visited Gibbard in Cambridge in December, just after the end of the Michaelmas term. Leading the way through dawdling crowds of tourists, his grey hair forming a sort of halo around his head, Gibbard turned to me: 'We're not very used to dealing with this sort of reaction,' he said. 'And while it's very nice to have the attention, I'm afraid this has produced quite a bad opinion of journalism in our group.'

After attending his local grammar school in Isleworth, Gibbard studied geology at the University of Sheffield and then decided to focus his research on the time period known as the Quaternary – the past 2.58 million years of Earth's history. For a geologist, a mere 2.58 million years is only semi-respectable, veering as it does into the territory of geography, which typically deals with the present and the more recent past. Indeed, his then head of department, a Carboniferous (359–299 million years ago) geologist, shook his head when he heard about Gibbard's choice. 'Oh dear,' he said. 'Where did we go wrong?'

As secretary-general of the ICS, Gibbard is one of the forty-two men and women – mostly men – in charge of the organisation of deep time. If the minds of eighteenth-century naturalists 'grew giddy', the twenty-first-century ICS stratigraphers have resolved to impose order. 'If I've made any kind of contribution, it's to bring to the attention of my colleagues that we have to formally define everything, because otherwise it just flaps around in the breeze,' Gibbard said.

In an effort to stop that, stratigraphers had been quietly getting on with dividing the 4.6 billion years of Earth history into units: ages, epochs, periods, eras and eons. Then, in the summer of 2018, they decided that around 4,250 years ago a cataclysmic upsetting of global weather patterns marked the beginning of an entirely new geological unit: the Meghalayan Age, a subdivision of the Holocene Epoch encompassing everything from that point up until the present day. The ICS sent out a press release, as normally happens in such situations, and then, as normally doesn't happen, everything went a bit crazy.

It is distinctly unusual for the spotlight of media attention to fall on stratigraphers. Those scientists concerned with the classification of layers of rock and ice are typically left alone to work on events that took place thousands, millions or – more likely – hundreds of millions of years before the beginning of human history. Even among geologists it's viewed as tedious work. Many universities no longer teach stratigraphy as a distinct discipline, preferring to fold it into more obviously 'practical' courses on, say, structural geology or palaeontology. 'If we tried to teach straight stratigraphy now, there would be a student revolt,' was how one lecturer put it to me. But following the formalisation of the subdivisions of the Holocene, angry op-eds appeared in the scientific and popular press. The new Age was a fix, a sham, a scam. At best an irrelevance – 'I'm never going to use the name, and I'm guessing most scientists will not,' the palaeoclimatologist Bill Ruddiman said in an *Atlantic* magazine interview. 'This is the age of geochemical dating … The community just doesn't care about these definitions.'[2] ('Geochemical dating' is a broad term that exploits the known rate of decay of unstable radioactive isotopes such as carbon-14.) At worst the new unit smacked of a sinister anti-green plot. 'The Meghalayan Age defines the "late Holocene" and the present day, but makes no mention of the human impact on the environment,' wrote Mark Maslin and Simon Lewis, climatologists at University College London. 'It seems like a small group of scientists – 40 at most – have pulled off a strange coup to downplay humans' impact on the environment.'[3]

The ICS found themselves on the defensive. 'Some disgraceful things have been said – quite disgraceful – by people who ought to know better,' Gibbard told me, sounding hurt. 'And the problem with a lot of what's been written is that it's by people who don't understand geology.'

*

Rocks are time made manifest. 'We can use radioactive decay to date things, but the only tangible evidence we have of time passing are the rocks themselves,' Gibbard had said.

Early one summer the palaeontologist and stratigrapher Jan Zalasiewicz took me to see an old railway cutting not far from his office at the University of Leicester. The exposed rocks dated from the early Jurassic (about 185 million years ago), when Leicester, and the rest of Britain, was covered by a shallow sea. Today that sea has become a pale, honeycomb-coloured limestone filled with the remains of the creatures that once thrived there – tightly spiralled ammonites, bullet-shaped belemnites, the curved gryphaea known colloquially as devil's toenails.

It was a warm day, and the cutting was filled with purple foxgloves. Birds that sounded like burbling water sang out of sight among the trees. Each centimetre of rock represented about a thousand years of time. Zalasiewicz, dressed in the traditional male geologist's outfit of hiking trousers, boots and fleece, showed me where, half-way along the cutting, the rock changed. The shell-rich limestone stopped, and above it was a dark blue-grey shale composed of thin, crumbly layers like stacked sheets of paper. There were no fossils in the shale. We were looking at the evidence of something known as the Toarcian Extinction Event, when an estimated 1.5 to 2.7 trillion tonnes of carbon were released into the atmosphere,[4] and the world warmed up by some 5°C.[5] The oceans became depleted of oxygen, and stagnant. A lot of things died out, which was why there were no fossils in the dark blue-grey layers. The presence of so much carbon was responsible for the dark colouring of the rock. 'It's one of the events used to compare with modern global warming,' Zalasiewicz said.

The transition between the two rock types, Zalasiewicz

explained, marked a transition between two worlds. A moment when the Earth changed for ever: one way of life ended and another began. Before the transition, the Earth was in a time unit known as the Pliensbachian Age. Afterwards, it was in the Toarcian.

As James Hutton knew, rocks are the Earth's history book – though one where many of the pages are missing, damaged, upside down, back to front and out of order. If you can learn to read them – to see a change of rock type and link that to an ancient climate event, say – then you can build up a history of the Earth and, eventually, a geological time-scale. Then you can start to say that this piece of rock in Greenland is the same age as this piece of rock in South America. That all these creatures in this part of the world went extinct at the same time as these creatures in that part of the world. You have a framework with which to understand everything that has been going on for the past 4.6 billion years.

'People may say it's dull, but without stratigraphy you can achieve nothing at all. Until you can order the rocks and link one area with the next, you can't start talking about things like evolution, palaeogeographic change, the migration of plants and animals,' Gibbard said.

The International Stratigraphic Chart is the official version of that time-scale.[6] A sort of esperanto, providing Earth scientists with a common language, it ensures that everyone is thinking about the same thing when someone says, for example, *Cryogenian*. 'It's the equivalent of the periodic table,' I was told. 'The one thing you'll find in every professional geologist's office.'

Within the chart, the geological units are hierarchical and nest inside one another like matryoshka dolls – from the smallest, an age, which typically covers a few million years, to the largest, eons, which are massive: all of geological time is divided into only three of these immense units. Some of the periods are familiar – Jurassic, Carboniferous – but a list of ages – Wordian, Roadian, Kungurian – sounds more like the roll-call for some intergalactic council meeting. Divisions between units are based on the occurrence of globally synchronous events that leave a changed world and leave a record of this change in the rocks and ice. For the Pleistocene/Holocene division this is the global warming event recorded as a sudden change in the composition of trapped air in the Greenland ice cores. For the Permian/Triassic, a time of rising temperatures, stagnating, acidifying oceans and catastrophic volcanism, it is a mass extinction event sometimes called (with geology's occasional flair for sounding like 1970s sci-fi) 'the Great Dying', when an unimaginable nine in ten marine species and seven in ten land species vanished. Life on Earth almost came to an end. The event is evidenced by changes in the carbon isotope ratio, layers of volcanic ash and dramatic excursions from the fossil record.[7]

Each unit also has its own colour. Right at the bottom of the chart the Hadean, a mysterious time when the surface of the Earth was covered with molten rock, is a livid purplish red. The early parts of the chart, when life existed predominantly in the oceans – the Cambrian, Ordovician and Silurian – are misty, watery greens and blues. For our own Holocene Epoch the geologists have for some reason chosen a slightly nondescript pinkish, brownish beige. Balanced at the top of the charts are the colour of a washed-out sticking plaster, the colour of calamine lotion or a wodge of well-chewed pink bubblegum. There is Plato and Aristotle, Marie Curie and Einstein, Mozart and Virginia Woolf, Confucius and the Buddha. And there beneath that little strip are all the other worlds we will never experience directly but that will come to us as fragments, as traces. The vanished worlds that came before our own.

The result of hundreds of years of painstaking labour by thousands of men and women, you might trace the beginning of the chart back to the work of Steno in the seventeenth century, and on through eighteenth-century workers such as Johann Gottlob Lehmann in Germany, who classified rocks in pursuit of tin, coal and other natural resources. The second time we met, Gibbard was working on a paper about the Italian geologist Giovanni Arduino, who in 1760 made one of the first attempts to put these rocks into an order based on time, dividing the different rocks he encountered in the Venetian and Tuscan regions into units: Primary, Secondary, Tertiary and Quaternary.[8] Today, in deference to Arduino, we call our own geological period the Quaternary.

In the early nineteenth century the English canal and railway surveyor William Smith recognised that fossils appearing within different layers of rocks might be used to identify the rocks and their relative age more precisely. His work was enthusiastically embraced, first by landowners hoping to find coal on their property, and then by the geologists. Known today as the principle of faunal succession, his technique went something like this: imagine a dinner party attended by Graham Greene and Britney Spears. When did it take place? We know that Greene became extinct in 1991 and that Spears first appeared in the record in 1981, so the dinner party had to have taken place within that ten-year block. This is what you can do with fossils in order to assign a rock to a geological period. If this ammonite is alive at the same time as this belemnite, then we must be in so-and-so period. Today radiometric dating can then sometimes be used to define the actual, as opposed to the relative, age of a rock and the fossils within it.

The first unit to be identified was the Silurian (444–419 million years ago), proposed in the early 1830s by the British geologist Roderick Impey Murchison, an ex-army officer whose rock-enthusiast wife, Charlotte Hugonin, directed him away from 'an idle life of fox hunting into his lifelong career as a geologist'.[9]

Following Murchison's work, the bulk of the periods were named during the nineteenth century, when geologists sketched out the basis of the geological time-scale we use today. Geology was arguably the pre-eminent fashionable science of the age, occupying a place in the public imagination today filled by neuroscience, AI or quantum physics, and referenced in the work of such towering cultural figures as Thomas Hardy, Alfred Tennyson and John Ruskin. Some periods are named after places – the Permian after Perm in Russia, the Devonian after the British county – some after their rocks – the Carboniferous, after the dark carbon-rich rocks we know as coal. Others – the Cambrian, Ordovician and Silurian – with a nod to the Victorian vogue for early British prehistory, are named after various Celtic tribes. At the top of the chart Charles Lyell attempted to impose some discipline with his series of Greek-derived names: Palaeogene – ancient birth; Neogene – new birth.

It was not until 1878, however, that the first International Geological Congress in Paris set as its objective the production of an internationally recognised standard stratigraphic scale.[10] And it is perhaps a reflection of the immensity of the task – in terms of both the science that needed to be done and the level of international co-operation – that such a scale or chart would not be finally agreed on for more than 100 years – the first ICS Global Stratigraphic Chart finally appearing in 1989.[11]

All of which is to say that, from a stratigrapher's perspective, altering the chart is kind of a big deal. Indeed, once a change is made there is a ten-year moratorium on any further alterations – 'Like the Tory Party,' Gibbard said. Were it decided that a mistake had occurred – the wrong start date for a period, the wrong prime minister – correcting it could take some time.

When it came to the Holocene, geologists had for years been talking about distinct periods of time within the epoch – such as, say, 'the early Holocene' – without ever specifying when, exactly, the early Holocene occurred. One person might say that the late Holocene was warm; another might say it was cold. Both people might be right because the term 'late Holocene' was used so inconsistently. The muddle had to stop. So back in 2009 Gibbard asked Mike Walker of the University of Wales to lead a team looking into the possible formal subdivision of the Holocene Epoch. At that point neither man realised just how heated things were about to get.

*

Geologists, the librarian and poet Michael McKimm had told me, are very keen on committees, and within the ICS each geological period has its own subcommission – a committee of scientists responsible for arranging, dividing and defining that segment of deep time. Sometimes things get territorial. There's a tendency to feel protective of *your* bit of deep time. 'It's human nature,' says Gibbard. When we put up borders around a piece of deep time, he suggests, the same impulse is operating as a person putting up a fence around their house to keep the neighbours out. As a former chair of the Subcommission on Quaternary Stratigraphy, he experienced this first-hand when he successfully argued for the formalisation of the Quaternary period with a base of 2.58 million years ago.[12] Achieving this necessitated the lowering of the base of the Pleistocene (an epoch within the newly formalised Quaternary) and the concomitant swiping of 0.7 million years off the top of the Neogene period – much to the disgust of the Neogene subcommission. ('I object to that term "swiping",' Gibbard says. 'It makes it sound like the Oklahoma land grab.')

John Marshall is the chair of the Subcommission on Devonian Stratigraphy. His office at Southampton University, where he is a professor of earth science, is so crammed full of books, papers, fossils and pieces of rock, that if one person wants to move around, the other must hover in the doorway. I sat down next to a lump of Fluorite that Marshall has had with him ever since he found it as a child in the Peak District near Macclesfield, where he grew up.

As an undergraduate, Marshall studied Natural Sciences at Cambridge – 'I realise now that a lot of people I met would have written me off as an oik,' he said. But in the science department he met other kindred spirits. 'The public-school people would tend to do the easier subjects: law, history and so on … The northern grammar schools provided many of the science students.' Today he coaches local state school pupils for the Oxbridge entrance exams.

Marshall likens the work of a stratigrapher to that of a historian. 'The skills which I need are grey data skills, the ability to make up things from incomplete data sets. We have quite a few people who come from physics and chemistry and they see [geology] as a science to be turned into physics or chemistry – but in stratigraphy you can't do that. You can't just measure an isotope.'

In the Subcommission, 'Essentially we look after Devonian time. It's like being a time lord, only you never actually go there.' On the Chart, the Devonian is a sort of toffee colour. It sits between the Silurian and the Carboniferous, in the Palaeozoic (old life) Era, in the Phanerozoic (visible life) Eon, and is dated from 419 to 359 million years ago.

The official marker of a transition between units is a 'Global Standard Section and Point' (GSSP), known as a 'golden spike' – in actuality a bronze disc hammered into the rocks at one representative geographical location on the planet where, for example, the Ordovician rocks end and the Silurian begin. (In this instance the location is a small, steep-sided valley called Dob's Linn, in the Scottish Borders.[13] The evidence for the transition is the first appearance in the fossil record of the graptolites *Parakidograptus acuminatus* and *Akidograptus ascensus*, examples of which can be found in the rocks at the site.) Where no GSSP can be identified – for the 4-billion-year-old Eoarchean, say, where there are few physical traces because so few rocks from that era remain undisturbed – then a radio-metrically dated Global Standard Stratigraphic Age (GSSA) is used instead. This gives a chronological point but not a physical reference.

International attitudes towards hosting GSSPs can be instructive. 'China has the GSSP for the Permian/Triassic,' Marshall told me, 'and it's in a public park with ornamental gardens and it's beautiful, but if you go to the Ordovician/Silurian one in Scotland its basically a wasteland. It's the corner of a rather nice little river, but there's nothing there – maybe a signpost on the road.'

Sometimes, countries become protective of their GSSP. 'One problem we have at the moment is Emsian time,' Marshall said. The Emsian is an age within the Lower Devonian Epoch. The GSSP marking its base is currently defined by a change in fossil lineage and sited in a section of rocks in the Zinzilban Gorge in Uzbekistan – 'Which we were delighted about. We are an international organisation and want GSSPs around the world.' But on closer inspection it became apparent that the GSSP had been placed far below the traditional Emsian base – it was too low down in the rocks. Following discussions, the Devonian stratigraphers began to work towards redefining the GSSP, but the task wasn't straightforward. 'The Uzbeks very much value having the GSSP in Uzbekistan; it gives them international importance,' Marshall said. It proved difficult to get back to re-study the section in Zinzilban Gorge. 'There were access problems with roads being blocked. It's in a border area, so it can be difficult to get permission.' In addition, the generation of geologists who defined the GSSP are no longer active, so finding the appropriate rocks for redefinition was difficult. These problems went on for a decade, until eventually the Devonian stratigraphers decided to reopen the redefinition to other sections outside the country. 'It's sad,' Marshall said. 'If they could turn up something at the right level in Uzbekistan then we'd take it tomorrow, but it looks as though we're going to have to move it.'

In the case of the Holocene, Walker suggested three new cuts in the rock record, producing units that – unusually for stratigraphy – are measured in thousands rather than millions of years, because the closer the stratigraphers come to our own 'historical' time, the greater the resolution and the more detailed the chart can be. Taken to its logical conclusion, I suppose, the divisions in deep time would get finer and finer until one day they were indistinguishable from the divisions of human time, and we, like Archbishop Ussher, could tell the month, the day of the week and the hour of the beginning of our world.

Walker's divisions were as follows. The early Holocene began around 11,700 years ago, as the last ice age ended. Stratigraphers call it the *Greenlandian* Age, because its golden spike has been placed in an ice core drilled from the Greenland ice sheet and maintained by Jørgen Peder Steffensen as part of the ice core collection at the University of Copenhagen. The middle Holocene began around 8,200 years ago, during a sudden and unexplained outbreak of cold temperatures in the northern hemisphere. It's named the Northgrippian Age, after the scientific expedition that found the ice core that houses its golden spike: the Northern Greenland Ice Core Project, or NorthGRIP. And the most controversial unit – the late Holocene – began about 4,250 years ago during a massive upsetting of global weather patterns. Some areas got hotter, some dryer, some wetter. Some places suffered an abrupt increase in acidification, other saw the arrival of neoglacial conditions. Across the globe agricultural societies were devastated, entire civilisations wiped out. Egypt's Old Kingdom and the Akkadian Empire in Mesopotamia collapsed. In the Indus Valley (an area covering parts of modern-day Pakistan and India) large, once flourishing cities such as Mohenjo-daro were abandoned. The GSSP for the Meghalayan is recorded in a stalagmite in a cave in Meghalaya, a state in north-eastern India, so the age is named the Meghalayan.[14]

Walker sent his proposal up the chain of command to the Subcommission of Quaternary Stratigraphy. One thing that pleases me about stratigraphy is the strange collision between the human impulse towards bureaucratic procedure – going on in some form or other since at least the development of Sumerian civilisation in 2,500 BC – and the very inhuman weirdness of deep time. The way stratigraphy uses numbers and a specific scientific vocabulary in an attempt to domesticate and rationalise the quite literal other-worldliness of the far past. Proposals relating to changes to geological units are no exception. After the Subcommission receives a proposal from the working group, a vote is held. If the Subcommission is in agreement, the proposal is passed up to the International Committee on Stratigraphy, which is made up of the titular members of the subcommissions. ('Its like being in the College of Cardinals,' Marshall said.) After that, the proposal finally arrives in front of the ultimate arbiters: the International Union of Geological Sciences (IUGS). Only at this point can it be formally ratified. Then, with the official backing of the IUGS, it at last takes its place – for a minimum of ten years – on the International Stratigraphic Chart.

*

On 8 March 2019, Mark Maslin, who, along with Simon Lewis has published a successful popular science book about the Anthropocene, tweeted: 'In your face #Holocene – because time for you to be written out of geological history!' One of the reasons so many people reacted angrily to the Meghalayan announcement was that they were waiting for a different decision from the stratigraphers: they wanted to know if we were now officially living in the Anthropocene.

Anthropos: 'human being'; *Kainos*: 'new'. In 2000 the atmospheric chemist Paul Crutzen and the biologist Eugene Stoermer outlined a radical new proposal: humankind had influenced the Earth's geology and ecology to the point of fashioning an entirely new geological epoch.[15] For the first time, we were the significant event that had changed the world and left a record behind in the rocks. Examples of human influence might include modified global cycles of carbon and nitrogen, rates of extinction well above background levels and a decisive warming of global temperatures. It was an awe-inspiring, terrifying idea. The Holocene Epoch was finished: we had entered the man-made age.

In 2008, at the instigation of Gibbard (at the time chair of the Subcommission on Quaternary Stratigraphy), Jan Zalasiewicz set up the Anthropocene Working Group to consider the stratigraphers' response. According to Zalasiewicz, 'In my personal opinion there's no question that we've entered the Anthropocene', but so far there has been no formal addition to the ICS chart. The investigation is ongoing.

Some cheerleaders for the Anthropocene concept, such as Maslin, are suspicious, and appear to have taken the Meghalayan as a direct attack on the Anthropocene. (This despite the fact that Zalasiewcz himself voted in favour of the new Holocene subdivisions.) Neither Maslin nor Lewis agreed to talk to me about their work, but in an article on The Conversation website Maslin argues that: 'The Anthropocene rewrites the human story, highlighting the need for planetary stewardship,' while 'The Meghalayan Age says the present is the same as the past.'[16]

'It doesn't say any such thing,' Gibbard said, when I put this to him. 'If he actually paid attention and read things carefully rather than just reacting emotionally, he would know that when you define a boundary such as the Meghalayan, you only define it by its base.' In stratigraphy, it's the base of whatever comes after the Meghalayan that will define the Meghalayan's top. And the unit at the very top of the chart must always be left open to ensure there are no gaps, hence the Meghalayan necessarily extending up to the present moment. So recognition of the Meghalayan does not negate the possibility of the Anthropocene. There's no reason why the Meghalayan and the Anthropocene can't one day co-exist on the chart.

But Maslin and Lewis also ask whether the entire Holocene concept is flawed. The Holocene began when the last ice age ended, but as Steffensen's cores show, during the last 2.58

million years – i.e., the Quaternary Period – there have in fact been a series of ice ages, or glacials, and warmer intervals or interglacials. We are living in an interglacial now – and substantially altering its climate, such that we have pushed back the arrival of the next ice age. And as no other interglacial has been given the rank of epoch, Maslin and Lewis write, singling out the Holocene for special treatment 'makes little sense geologically'.[17] Perhaps the earlier stratigraphers made a mistake. Perhaps it is time for the Holocene to be retired. They argue that the 'Geologists should set up a new official multi-disciplinary IUGS commission with the remit of proposing the classification and definition of this geological time we live in, to happen within two years.' The Anthropocene Working Group – which is to some extent multi-disciplinary – has several times invited Maslin to join them. So far this offer has been declined.

Other scientists, meanwhile, have expressed dissatisfaction with the way the Meghalayan boundary was fixed in time – and those arguments go right to the heart of what makes the discipline of geology so peculiar.

This is a 'paleoclimate white whale', a palaeoclimatologist complained in an article in *Science* magazine.[18] 'The ICS mistakenly lumped together evidence of other drought and wet periods, sometimes centuries away from the 4,200-year event, to mark the beginning of the Meghalayan.' The same article pointed out that an unpublished analysis of other Meghalayan stalactites by scientists from Xi'an Jiaotong University in China found a steady weakening of the monsoon over more than 600 years, rather than a sudden drought 4,200 years ago. Gayatri Kathayat, who led the research, did find evidence that 4,000 years ago there were several decades-long droughts that could be said to match the golden spike 'to an extent' but not entirely.[19]

Two hundred years might seem like a large margin of error, but for stratigraphers, especially those working before the Quaternary, getting to within a million years of an event is seen as impressively accurate. Furthermore, geological boundaries are diffuse. A transition between geological units is not comparable to the flicking of a switch. Rather, it is a slow and messy process that can stretch over hundreds and thousands of years, and so the fact that the climatic event appeared to cover multiple centuries wasn't necessarily an issue for the stratigraphers.

It comes down, perhaps, to an issue of perspective. Where some Earth scientists working on a human time-scale saw inaccuracy, the stratigraphers, attuned to the workings of deep time, saw precision.

6

THE DEMON IN THE HILLS

I had climbed the hillside to think about movement, but everything around me suggested stillness. No breeze stirred the vegetation. The roads were empty, as though the local population had already abandoned this place. In the distance bright green vines glowed in the afternoon heat.

From the campus of California State University, San Bernardino (CSUSB), to the foot of the San Bernardino Mountains is around three-quarters of a mile. Dusty scrubland slopes gently uphill, dotted with strong-smelling sagebrush and stringy wild sunflowers. Closer to the mountains, small trees grow along a strip of fertile land covered with bright green grape vines. There are a handful of old farmsteads, but any new building has been forbidden since the early 1970s. The owners of the farmsteads are mostly unable to sell their properties because no one would be able to secure a mortgage to buy here.

Passing signs that warned of rattlesnakes and mountain lions, I cut left on to a sandy path. Beyond the mountains lay the baking, arid sweep of the Mojave Desert. Los Angeles was around sixty miles west. Away in the scrubby bushes a bird, unseen, was calling – a shrill peep-peep-peep of alarm.

Somewhere close by was the reason these farmsteads were being slowly abandoned: the San Andreas Fault.

*

Our apparently stable world is in constant motion. 'If you wait long enough, everything flows,' says University of California, Los Angeles (UCLA), geodynamicist Carolina Lithgow-Bertelloni. 'Over very, very long time-scales even the rocks flow, just like they were water in a boiling pot.'

Lithgow-Bertelloni's work builds on a revolutionary theory that was developed in the 1960s and changed for ever our understanding of the Earth: plate tectonics. 'It's difficult to overstate its importance,' she said. 'Before plate tectonics we had centuries of observations of rocks and rock types and fossils, but there was no framework with which to understand why they were the way they were.' In the film *Antarctica: Ice and Sky* the venerable French glaciologist Claude Lorius describes the '60s as a sort of golden age before our long reckoning with global warming and human destructiveness began. The moment in history when man travelled to the moon, and scientists believed that they were about to crack open all the secrets of the natural world. Among their number were the men and women racing to develop the theory of plate tectonics.

The formulation of the theory made geology a 'respectable' science at last. Until that point it had been largely observational and essentially an exercise in collecting apparently disparate facts: the sort of thing Ernest Rutherford had in mind when he (allegedly) opined that 'all science is either physics or stamp collecting'. Plate tectonics provided geology with a grand theory – the equivalent of the arrival of Darwinism in biology, or quantum mechanics in physics. It's the scaffolding on which we've built our present understanding of the Earth. 'There are still plenty of things to discover, but I doubt anything else will happen in terms of such magnitude,' says University of Durham geodynamicist Philip Heron.

According to the theory, the Earth's surface is composed of a shifting mosaic of great slabs, around 125 km thick, of uppermost mantle and crust.[1] There are around seven major large slabs or plates and perhaps eight smaller ones – opinion varies on the precise number – and they move around the globe, gliding atop a sort of conveyor belt of hot, weak rock, carrying the continents and oceans on their backs. Where they meet at plate boundaries they pull apart, scrape against one another or collide together like a car crash happening in slow motion. One plate knocks into another, and the land buckles like a crumpling bonnet, and hills and mountain ranges are created. Sometimes the plates dive underneath one another, pulled and pushed down into the red-hot depths of the mantle where the rocks are melted and recycled. These interactions between the plates have helped geologists explain why our world is the way it is: from the shape of the land, and how that has changed through time, to why volcanoes erupt and earthquakes shake (the result of movements within the Earth's crust generated by colliding plates) and the distribution of plants and animals around the globe.

In a human lifetime of, say, eighty years, a plate might travel between 2 and 4 m. My flat in south London, for example, was built in 1980. Since then, because of plate tectonics, its position on the surface of the globe, measured in latitude and longitude, will have moved about 1 m to the east, travelling at a speed of around 2.5 cm a year – roughly the rate of growth of your fingernail. A velocity of 2.5 cm a year puts us within the realm of deep time, of scales too large or, conversely, too slow, to register on the human senses. At speeds like this, on a deep-time-scale, whole continents with all their flora and fauna can shift thousands of miles apart. Oceans can appear or disappear.

Plate tectonics changed everything, yet in many ways we are still near the beginning of this journey. As a theory, it is only around fifty years old. In 1983, when John McPhee published *In Suspect Terrain*, he was still able to find respectable scientists who took issue with it.[2] 'It seems incredible to me to think that when Barack Obama was in school, he would not have been taught that the continents move over time,' Heron told me, shaking his head.

Depending on who you speak to – depending on age, inclination, nationality, friendships and rivalries, depending on whether you place more emphasis on the person who had the hunch, the person who found the proof, the person who worked out the formula or the person who synthesised the findings – you can tell the story of plate tectonics in many different ways. This means that in recent years there have been several separate celebrations of its fiftieth anniversary. In 2013 the influential journal *Nature* marked the fiftieth anniversary of the publication of a paper by British geologists Frederick Vine and Drummond Matthews;[3] in 2016 Columbia University celebrated a paper from 1966 by their researchers Walter Pitman and James Heirtzler;[4] in 2017 the Geological Society of London commemorated a paper written in 1967 by two more British geologists, Dan McKenzie and Robert L. Parker;[5] and in 2018 the Collège de France celebrated a paper from 1968 by their former professor Xavier Le Pichon.[6]

Still, many accounts begin with Alfred Wegener. Wegener was a German meteorologist, geophysicist and polar researcher, but today he is most often remembered as the author of *The Origin of Continents and Oceans* (1915), in which he outlined his theory of continental drift.[7] For many years people had noted a strange similarity between the west coast of Africa and the east coast of South America. It was as though a china plate had been broken in two and the separate halves moved apart, the South Atlantic Ocean inserted in between. It was almost as though they had once been joined together. Wegener proposed that all the Earth's continents had once formed one 'super-continent', which he named Pangaea (from the Ancient Greek *pan*, 'all, entire, whole', and *Gaia*, 'Mother Earth, land'). Over time Pangaea had broken apart because the continents were not fixed in position but instead moved or 'drifted' across the surface of the globe.

In his lifetime Wegener's ideas were widely mocked both in Germany and abroad. Critics scorned what they called his 'delirious ravings'.[8] For decades afterwards, young geologists were warned that any hint of interest in continental drift would doom their careers. Crucially, although there was plenty of evidence to support Wegener's theory, from the shapes of the continents to the spread of flora and fauna, he couldn't explain how the plates moved. When he died between supply camps on an ill-fated expedition to the Greenland ice sheet in 1930, Wegener was just past fifty. Had he survived the expedition, it is not inconceivable that he would have lived long enough to see his work hailed as the beginning of a scientific revolution.

The discovery that finally brought Wegener's theory in from the cold was called sea-floor spreading. The geophysicist and marine geologist Tanya Atwater, now professor emeritus at the University of California, Santa Barbara, was twenty-five when she arrived at the Scripps Institution of Oceanography in January 1967 to find the place in chaos following a talk on sea-floor spreading by Fred Vine, one of the authors of the 1963 paper. 'Apparently the whole institution attended the talk, most of the scientists going in as fixists, all coming out as continental drifters,' she later wrote. 'In the first meeting of my first class, marine geology, Professor Bill Menard forgot to tell us any of the usual class preliminaries, just launched into raptures about this "wonderful new idea", scribbling all over the blackboard.'[9]

Sea-floor spreading occurs when hot material from the mantle wells up between two plates on the sea floor, pushing them apart and cooling on to the edges of the plates. This discovery at last provided the mechanism missing from Wegener's theory. This was the motor causing the continents to drift. 'I often couldn't sleep at night, my head was so abuzz with geo-possibilities and implications', Atwater remembers. 'It is a wondrous thing to have the random facts in one's head suddenly fall into the slots of an orderly framework.'

Popular accounts of scientific discoveries are often biased towards the single great discovery by the single great person – much easier to celebrate in an anniversary and far more romantic. Real life is mostly not like that. Scientific discoveries rarely spring into existence fully formed on a certain day. More often they are the result of many years' thinking and many different people coming together in the right places, at the right times. The story of plate tectonics follows that pattern. In early 1967, four years after Vine and Matthews's paper, British post-doctoral researcher Dan McKenzie was at the American Geophysical Union Spring Meeting and due to attend a talk by the Princeton lecturer Jacob Morgan. After scanning the abstract, McKenzie decided to skip it. Morgan, meanwhile, proceeded to veer wildly from his published abstract to present what was effectively the first description of plate tectonics, drawing on the work of Vine and Matthews, Pitman and others to explain how tectonic plates move around the surface of the globe. It was eerily similar to the work McKenzie was himself pursuing.

In November of that year McKenzie and Parker sent their paper setting out the mathematical principle of the movement of tectonic plates to *Nature*. An accompanying letter urged that 'unnecessary delays' in publication 'be avoided'. In December, McKenzie, having learned that Morgan had also written a paper on the new theory, wrote again wondering instead if publication *could* be delayed, so that both papers would appear at the same time. The editors replied that the issue had already been typeset and publication would go ahead. Morgan's own paper was held up in the peer review process and finally came out in 1968.[10]

*

To see the effects of this inexorable motion, just go for a walk on the South Downs (the African plate crashing into the Eurasian plate). It's much harder to get a visceral sense of the plates themselves. They are just too large; they move too slowly. I started thinking of the plates in terms of islands, and then I thought of the myth of the world carried on the back of a giant turtle swimming through a cosmic ocean. From the middle of an island or the back of a giant turtle, you'd have to make your way to the very edge to find direct proof of what it was you were living on. I decided that I wanted to visit a plate boundary. Many are under water, and there are few places in the world where they can be seen at the surface as an obvious trace. One of these is marked by the San Andreas Fault, where North America scrapes against the Pacific plate.

All day and night the slow, deep-time motion of the plates whispers through the faults that run alongside the San Bernardino Mountains, through the lonely vastnesses of the Mojave Desert, underneath the cinemas and theatres of Hollywood

Boulevard, down underneath the restaurants and bars of Santa Monica and the Venice Beach sidewalk, underneath suburban backyards in the town of Hollister and past the San Francisco General Hospital.

A fault is created when stresses and strains accumulate within the rocks far below our surface world. Under immense pressures from the moving plates, the rocks begin to bend until one day the fault ruptures, the rocks breaking violently apart. We call this an earthquake. The ever-looming threat of earthquakes on the San Andreas Fault is one of the consequences of living near this particular plate boundary.

Before I set off to look for the fault, I went to see Joan Fryxell, a professor of geology at CSUSB. From the window of her office you can see about three-quarters of a mile uphill to where those bright green trees and grape vines mark the line of the fault. (The greenness of the vegetation shows that it is growing on a spring line, a more benign sign of the disruption of the landscape at the plate boundary.) For someone who studies earthquakes the location is either extremely fortunate or very, very bad. One United States Geological Survey (USGS) seismologist told me that she hoped her career would be a dull one. Emphatically she would *not* like an office next to the San Andreas. Fryxell is of the other persuasion.

When I asked her what everyone asks her – *Is the San Andreas ready to rupture?* – she said: 'If it happened right now, I wouldn't be surprised. If it waited ten years I wouldn't be surprised. If it waited until after I moved away or died I'd be very disappointed. It's literally a once-in-a-lifetime thing, and it's in the window of being ready to go. I'd kind of hate to miss it.'

Fryxell has short grey hair, and a long, quizzical face. She told me that earthquake researchers are limited by what is otherwise a great boon – the relative infrequency of major earthquakes. For example, it wasn't until he was able to study the great 1906 San Francisco quake that the geologist H. F. Reid established a clear dynamic relationship between faults and earthquakes. Publishing his report in 1910, he called his new theory 'Elastic Rebound', and it remains one of the foundations of modern tectonic studies. Faults remain locked as forces from the tectonic plates create pressure on and around them. This leads to a build-up of energy, which is released when the fault lurches abruptly. Broadly speaking, the longer the fault, the greater the build-up of energy and the potential size of the earthquake. (In recent years this calculation has been slightly complicated by the fact that rupture events can apparently also 'jump' between unconnected faults.) For around fifty years people thought the San Andreas Fault was a lot shorter, and therefore less dangerous, than it turned out to be.

The San Andreas is what is known as a strike-slip fault. Place your palms together with your fingers pointing directly ahead. Now slide your right hand forwards. This is the way a strike-slip fault moves. To map one of these faults, look for a rock formation – such as a specific granite or sandstone – that appears in different places on both sides of the fault. If you can show that the two parts of the rock formation were once next to one another, then you know that they must have been moved apart by the motion of one side of the fault slipping forwards. Next, you can calculate that if the two halves of the rock formation are, say, 100 miles apart, then the fault is at least 100 miles long. In the early 1900s, when the San Andreas Fault was first being mapped, people looked for matching rock formations on either side of the fault and found nothing. The reason? No one was looking over long enough distances. No one expected it to be quite so large.

It wasn't until the mid-1950s, long after the establishment

of Los Angeles and San Francisco as major cities, that the full extent of the danger was understood. The fault turned out to be more than 800 miles in length[11] – stretching from the Salton Sea in Imperial County to Cape Mendocino in Humboldt county – and around 10 miles deep, giving it an upper magnitude limit of 8.0.[12] (The size of an earthquake is most commonly measured scientifically using the moment magnitude scale. A magnitude 2.5 or less is not typically felt. Magnitude 4 is considered minor, and 6 is moderate – property damage can be expected. Magnitude 7 is considered strong, with an energy release greater than at Hiroshima. Loss of life is likely. Anything measuring 8 and over is classified as great: for context, any community near the epicentre of a great quake can expect to be totally destroyed. The San Andreas Fault's 8.0 upper magnitude limit depends on the entire 800 miles of the fault rupturing – an event with a 7 per cent probability of occurring within the next thirty years. The statistically more likely scenario is a magnitude 7.0 quake, which would come from the rupture of only a segment and has a 75 per cent probability of occurring during the same time period.[13])

'When it was first suggested it was a very large strike-slip fault, the geologists were seen as a wild-eyed radicals,' Fryxell said. By the time everyone had accepted the truth, it was too late to move California's major financial, cultural and residential hubs to safer locations. 'And whether it's ready to go right now is a question only known to itself,' she said. Then she shook her head and laughed: 'I anthropomorphise the faults and I know that's ridiculous. They don't have consciousness; they're not doing it to be mean. It's not like it's a demon down there.'

There may be no demon, but there was something brooding about the quiet of the still afternoon heat. As I tried to find the fault I had the sense of the landscape around me holding its breath. So easy to spot from a distance, as you get closer the fault begins to disappear. The line of bright green vegetation that had been so distinct now blended into its surroundings. Underfoot there was little to suggest anything unusual. Only a small, steep rise of about a metre in the path. Some pale soil that looked a little churned. There is no obvious fracture that you can stand astride, one foot on the North American plate, one foot on the Pacific.

Disappointed, I unrolled the map I was carrying – a gift from Joan Fryxell. It wasn't a hiking map but a geological one designed to be pinned on a wall, and was large and unwieldy. But looking at the map, you could see at once that something was up with the landscape.

The map was a quadrangle, bisected roughly in the middle by a black diagonal line. In the top half the contour lines were all bunched together, centred like whirlpools on a series of points indicating mountain peaks. The colours ranged from purplish pink through rusty orange to dark mustard yellow and pale blue. The bottom half of the map was quite different, much calmer. Few contour lines, large blocks of colour ranging from white to a bright pastel yellow, moving through shades of cream and cowslip, like a decorator's colour chart.

The mountains and the plain, that was clear enough. Now I began to look more carefully. As I looked, moving between the map and the mountains, the world took on a subtly different sheen. I was seeing not just the landscape around me but the story of that landscape. Trying to see, like a geologist, the landscape as an entity in time as well as space.

Three years after Tanya Atwater arrived at Scripps, she

became one of the first people to use the new theory of plate tectonics to answer a question about a specific landscape – and the landscape she had in mind was the San Andreas Fault. The late David L. Jones of the University of California, Berkeley, said of her work: 'That was the first application of plate tectonics to a real setting, and she was able to show people who had been fussing with the San Andreas Fault all of their lives that they were completely missing the story. It was a marvellous paper, and that's what convinced us that plate tectonics was the way to go.'[14]

The San Andreas Fault, Atwater surmised, came about because around 28 million years ago the Pacific plate, travelling north-west, came into contact with the North American plate.[15] That explained the difference between the two sides of the fault. The flatter, sandy, scrubby land is on the Pacific plate, and is mostly composed of Holocene deposits – only thousands of years old. The youngest are what geologists describe as 'slightly consolidated', meaning that they are only starting out on the journey to becoming 'proper' rocks. The mountains, which are whiter in colour, are on the North American plate and around 160–70 million years older. The predominant rocks are metamorphic and igneous, from the Cretaceous and the Jurassic: schist, gneiss and granodiorite.

Here two different worlds, two different temporalities, came together. You could tell, even as a non-geologist, that these were different landscapes, that they had lived through different things. The metamorphic rocks on the North American plate, for example, probably began life as mud and sand, laid down on a sea floor then slowly buried, compressed and heated before being raised up by the forces of the plates to re-emerge into the light as schist and gneiss. Some of these rocks would have been walked on by dinosaurs underneath an ancient Cretaceous sun. The Pacific side was newer, scrappier. These deposits had mostly arrived as pebbles, gravel and sand via landslides and rivers. They had not been buried and had not had time to consolidate, to bake down, to take on the ancient, weathered aspect of the older landscape.

When the North American and Pacific plates came together, the Pacific plate, travelling north-west, began to grind against the North American plate as it passed. This means that, as you read this, the rim of California is being hauled off towards Alaska. It isn't being hauled off steadily, but in a series of jerks that tear and shear at the rock, causing the network of faults that make up the San Andreas Fault zone. Currently the average rate of motion is around 46 mm a year.[16] Assuming this continues, Los Angeles and San Francisco will be next-door neighbours in around 15 million years' time.

I sat down and drank some water, breathed in the sagebrush, with its incongruous scent of Sunday roast. The bird I'd heard earlier had given up or retreated. Clouds of small black flies appeared, disappeared. I thought about shifting, drifting rock. The North American and Eurasian plate boundary in the Atlantic Ocean is in the process of widening, pushing North America and Europe further apart and making the flight from London to New York a few centimetres longer each year. The African plate, drifting northwards, is closing up the Mediterranean Sea. One day you will be able to walk in a straight line from Morocco to Madrid. In the Great Rift Valley – another on-land boundary – the African plate is being ripped slowly apart. One day a new ocean will be born there.

Compared with scientists in the 1960s and '70s, today's geodynamicists have more advanced technology and an additional fifty years' worth of data, all of which brings new insights. Plate motion, for example, can now be measured via GPS. 'I find it quite wonderful,' the science historian Naomi Oreskes has written, 'that a system developed to track missiles and other fast-moving man-made objects is equally useful for monitoring the slow, stately drift of the plates.' [17] Thanks to GPS, we now know that rock deformation actually occurs not just at plate boundaries but within the interiors of the plates. These are the sorts of questions that occupy Carolina Lithgow-Bertelloni: 'The relationships between the interior forces and the plates themselves; the strength of the plates; why plate motions stay constant for long periods of time …'

Elsewhere, seismologists track energy waves from earth-quakes to calculate the origin of a quake. 'It wasn't as if everyone before 1960 didn't know what they were doing. It's just that the data wasn't there,' Heron said. He showed me a map from 1960 marked with three earthquakes that occurred in the west of South America. Then he showed me a map of the same region showing those earthquakes plus all the other ones recorded between 1960 and 2018. On the new map you could see that the earthquakes formed a sort of horseshoe shape – the shape, in fact, of the Nazca plate disappearing underneath South America. As the plate lurches downwards into the hot mantle, earthquakes occur. The shape the earthquakes make on the map tells us what is happening to the plate, but there's no way you could have deduced that from the paucity of the data in 1960.

Some scientists have speculated that it was the collision of plates, and the resulting break up of rocks, that released crucial nutrients at key moments of biological evolution, such as the 'Cambrian explosion' of about 540 million years ago, when the ancestors of modern life forms first appeared. It is increasingly obvious that 'you need plate tectonics to sustain life,' Aubrey Zerkle of the University of St Andrews has said. 'If there wasn't a way of recycling material between mantle and crust, all these elements that are crucial to life, like carbon, nitrogen, phosphorus, and oxygen, would get tied up in rocks and stay there.' And crucially, the conveyor belt of plate tectonics helps prevent the dangerous build-up of carbon dioxide in the atmosphere by taking carbon-loaded rocks and pulling and pushing them underground to melt back into the mantle. Plate tectonics helps us to breathe.[18]

*

In her office at CSUSB, Fryxell is more concerned with the dangers that living on a plate boundary brings. She exists in a state of constant earthquake alertness. The bookshelves have lips, so that the books won't slide out if a quake hits, and her rock collection is stowed safely on the bottom shelf. The blinds are always lowered against the danger of breaking glass. Her students, by contrast, appear largely untroubled: 'Landers [a magnitude 7.3 earthquake] was widely felt through here, but that was 1992 and most of these kids weren't around. They haven't experienced a major earthquake even from somewhat afar, so I don't think it's on their radar in any sort of emotional sense. They don't know what it feels like, so they don't know how scary it can be.'

By mapping faults and studying seismic activity, seismologists often (though not always) have a good idea of the areas at threat from earthquakes. What they can't do is predict when such an event will happen. According to the USGS web site: 'Neither the USGS nor any other scientists have ever predicted a major earthquake. We do not know how, and we do not expect to know how any time in the foreseeable future.'[19] Like trying to figure out which straw will break the camel's back, to be able to predict an earthquake you'd need to be able to predict when exactly the stresses and strains building up in a particular section of rock will reach breaking point, shattering the granite or basalt or sandstone and unleashing rolling waves of energy through the landscape. 'And we cannot do that,' Fryxell said.

Individual earthquakes cannot be predicted, but scientists can say things about the statistical probability of an earthquake occurring within a certain (deep) time frame. By studying the sites of old earthquakes across the San Andreas, it can be seen that large earthquakes have occurred on the fault on average every 100 to 200 years. (Because the San Andreas is a fault system, not all sections have the same pattern. Near Palm Springs, for example, the frequency is every 200–300 years, whereas Wrightwood in San Bernardino County can expect a quake every 100 years or so.) Fryxell believes that the San Andreas Fault is 'in the window of being ready to go' because, as USGS seismologist Kate Scharer explained to me: 'Right now, everywhere you look the average time between earthquakes has been, or is about to be, exceeded. If it's typically about every 100 to 200 years, then depending on where you are it's been 150 to 200 years. So yes, it's a little intimidating.'

I asked Fryxell what would happen if a major earthquake were to strike while we were sitting in her office. There might be foreshocks, she said, but not necessarily. Otherwise the first thing to reach us would be the P (primary) waves, which if felt, are like a punch, like someone just slammed a door in the house. Between a few seconds and a few tens of seconds later the S (secondary) waves arrive, followed by the surface waves. These are the waves that do most of the shaking – levelling older buildings not built to withstand quakes, disrupting roads and severing electric, telephone and water lines. In Fryxell's office – which has been built to withstand an earthquake – we would get underneath the desk and hold on while the building bucked and rippled around us. Outside the window, as the fault slipped, we would see the mountains shoot three to six metres to the right, propelled by the accumulation of millions of years of strain. 'So I'm supposed to be ducking and covering, but I'd want to be looking out the window,' Fryxell said with a sigh.

Great cracks would appear near the fault, but since Fryxell's office was built after the Alquist Priolo Special Studies Zone Act of 1972, we would not be at direct risk from ground rupture because we are around three-quarters of a mile distant from the fault.[20] The Act forbids building on top of an active fault except where geological investigations can show that the fault presents no hazard to the proposed structure. This is why the land around the farmsteads up the hillside will not be further developed. It's also why, in fault-strewn cities such as San Bernardino, children's playgrounds, parks and other public spaces are sometimes placed directly on top of a fault – making use of land that lies within the city but cannot be built on, and providing interesting possibilities, you can't help thinking, for disaster movies.

What really worries Fryxell is what happens after the initial quake. One pervasive myth is that southern California could shear off the mainland, becoming an island. That's not going to happen, but in the months following a major quake, the land south of the San Gabriel and San Bernardino mountains – including Los Angeles, home to around 4 million people – might become something like an island, cut off from the resources of the mainland.

The USGS has led a disaster-planning exercise exploring what could happen if a magnitude 7.8 earthquake hit the southernmost 300 km of the San Andreas fault. Immediately after the quake a major danger would be the hundreds of fires expected to start up. With roads blocked and the water system damaged, emergency personnel would struggle to reach them and smaller fires could merge into larger ones, taking out whole sections of the city. The lines that bring water, electricity and gas to Los Angeles all cross the fault. Fixing them could take months. Though most modern buildings would survive the shaking, many would be rendered structurally unusable. Older ones would be destroyed. In the days following the quake a series of aftershocks would continue to shake the broken, traumatised city, bringing further damage to buildings and hampering rescue and relief work.[21]

Overall, researchers have estimated, such a quake would probably cause some 1,800 deaths, 50,000 injuries severe enough to require emergency room care, losses of buildings totalling $33 billion and over $200 billion of economic losses.[22] In the aftermath, without a functioning infrastructure, the local economy could easily collapse. Finally the people, bundling up their belongings as best they could, would abandon the shattered, blackened city, leaving Los Angeles to the rats and the cockroaches and the abandoned pets scavenging in the empty streets. The next big quake – what locals refer to with a mixture of fear and bravado as 'the big one' – could be the end of life as we know it on America's west coast.

*

Californians respond to this existential threat in different ways. At Universal Studios Hollywood you can take an earthquake simulation ride. (In a nod to more delicate sensibilities the ride was temporarily closed following the Northridge quake in 1994, which registered at magnitude of 6.7 and had an estimated death toll of fifty-seven.[23]) In 2015 Warner Bros released a film starring Dwayne Johnson and called simply *San Andreas*, and Angelenos happily paid out for tickets to see their city get torn apart and Kylie Minogue fall off a skyscraper. (The buildings collapsed pretty authentically, one USGS employee told me, but the San Andreas would never cause the film's climactic tsunami in San Francisco Bay.)

For some, the threat of a quake is just a backdrop, long subsumed under more insistently immediate concerns. 'It's the wildfires I worry about,' a taxi driver told me. But David Ulin, a Los Angeles resident and author of *The Myth of Solid Ground*, writes that:

> Once you've been through the cycle you never lose the edge of awareness, that anticipation of the quake to come. For years after leaving San Francisco I would tense up at the rumble of the New York City subway or feel a strong gust of wind shake the walls of a house [...] and experience what I came to recognise as a muscular memory, a clenching of both body and imagination.[24]

Others become fascinated by the faults themselves. Back in 2000, USGS seismologist Susan Hough was driving through the town of San Fernando, north of Santa Monica, scouring the streets for evidence of the San Fernando or Sylmar quake of 1971 – a magnitude 6.7 earthquake that caused over $500 million of damage and killed sixty-five.[25] 'Time and the dark forces of road repair and urban development have softened these once-crisp features,' she wrote disappointedly in *Finding Fault in California,* a sort of I-Spy guide for earthquake tourists.[26] But she found something interesting in the parking lot of the McDonald's restaurant on Glenoaks Avenue. 'I call this my burgers and shakes location,' she said, showing me the picture. The car park is on two levels, divided in the middle by a steep, grassy, green lawn and flowerbed. That green slope exists because, when the San Fernando fault zone ruptured in 1971, the land here was displaced, creating this slope or, the technical term, 'scarp'. The owners of the lot had decided to landscape the scarp rather than bother levelling it. Today the site has become a popular destination point on earthquake tours. Secular pilgrims seeking evidence of the immense powers of the tectonic plates among the everyday – in the streets and buildings of suburban housing developments and the car parks of popular fast-food restaurants.

Hough is a slight, quietly spoken woman. She drives manual in a city full of automatics, with a bobbing Hula dancer on the dashboard, a souvenir from a recent work trip to Hawaii ('the ultimate American kitsch – my kids hate it'). For the fault book she spent many weekends alone criss-crossing the state, from the centre of Hollywood to remote parts of the Carrizo Plain. 'Those roads were kind of sketchy,' she remembers. 'They would have been bad places to break down.' On the day we met, we drove into Hollywood to look for faults. Directly above us the sky was blue, shading to a smoggy brown in the distance. Purple flowering jacaranda thronged along the highway and up into the hills. As a seismologist, Hough's official life is mathematical and computational in nature, typically involving analysis of earthquake waves recorded by seismometers. 'Certainly through the first part of my career you could study earthquakes without thinking too much about faults other than as a source of energy. I knew about the faults, but I didn't have a good sense of where they were,' she said.

Living in California eventually changed her thinking. 'I realised that I spent my days generating maps on the computer, and when I'm not working I spend a lot of time taking my kids out etc. So I realised that I had two separate maps in my head.' A map of the faults and a map of the landscape. 'I wanted to bring those maps together.'

Hollywood Boulevard had the usual roving hustlers dressed as superheroes and comic book characters. Hough pointed out the historic Chinese Theatre, venue of the 1977 *Star Wars* première. We parked on Vine, just north of the famous Hollywood and Vine intersection, alongside the stars of John, Paul, George and Ringo. Hough, wearing a floppy canvas field hat and hiking trousers – as though we were headed out of rather than into town – gestured up the road towards the round tower of the Capitol Records Building, where Frank Sinatra, the Beach Boys and Nat 'King' Cole once recorded. Just in front of the building the road rose steeply uphill. 'There's the scarp,' she said.

This was the scarp of the Hollywood fault, which runs for around nine miles along the northern edge of the Los Angeles Basin. So far in recorded history it has never produced a significant earthquake, but if the fault eventually ruptures, its size and the local geology could produce a magnitude 7.5 quake with an epicentre in the middle of Hollywood. I tried and failed to really imagine walking to work every day over a potentially lethal piece of ground. 'I'll talk to people about the fault, and I'll ask them if they've driven on this stretch of road, and they'll say, I always wondered why that funny little hill was there,' Hough said.

If you know where – and how – to look, southern California is full of evidence of moving, twisting, bending rock many metres below ground level. Full of the evidence of slow deep-time processes expressing themselves in our busy surface world. Further north, towards San Francisco, favourite fault-finders' destinations are the cities of Hollister and Haywood, and the Haywood Fault. 'It's one of the best places to see creep,' Hough explained. Creep occurs when, instead of moving only during major earthquakes, the usual pattern for faults, they continuously 'creep'. Hollister is being very slowly ripped in two, mostly along a narrow zone running right through the middle of the town. In Hollister, Fryxell took a photograph of 359 Locust Street, where steady creep has shifted the path to the front door, so that it no longer leads to the centre of the porch steps. Elsewhere the low concrete wall outside a neighbouring house has been smoothly bent into a curve, as though the concrete were as flexible as Plasticine.

Driving back to the USGS field office, Hough told me about another way the evidence of faults has been recorded. You can't help noticing, she said, that many of the State's petroglyph sites are also places of high seismic activity. Coso, for example, which boasts one of the biggest concentrations of rock art anywhere in California, is similarly well known for earthquake and volcanic activity. As a scientist, Hough is wary of assuming that correlation is causation. As a fault-finder, she can't help wondering. Explicit earthquake-related legends do not abound in the oral histories of California tribes, but there are hints. Wavy lines drawn within figures of shamans in Little Petroglyph Canyon, for example, may imply unrest in the earth. Several tribes have legends describing caverns that moan and roar, and canyons haunted by malevolent spirits.

In her book Hough writes:

Considering the extent to which earthquakes are etched into the collective psyche of present-day Californians, perhaps it would only be surprising if the surviving written expressions of the state's earliest human inhabitants did not reflect the region's tumultuous natural environment.

California has always been Earthquake Country; Californians have always found, and will always find, ways to come to grips with this.[27]

*

If scientists living in earthquake territory sometimes fall into patterns of thinking that are not quite respectable, non-scientists are prone to outright superstition. A yellowing sky, an anxious dog, a still-warm evening, a feeling of dread, a sharp headache or aching hip, all presage some stirring in the Earth's lower depths.

'Earthquake weather is one I hear a lot,' Fryxell told me. 'Psychologically it makes sense. It gives you the illusion that you have some handle over what is going to happen.' Looking for something to act as a warning before the next earthquake, you think back to the weather conditions just before the last quake. Might they indicate trouble brewing? 'But if, as one of my friends swears, earthquake weather is warm and dry and still, then we would never have earthquakes in the Himalayas for example, or in any rain forest, or in any place cold. And of course the people who live in those areas will have their idea of what earthquake weather is, and it's whatever is a typical pattern for that area.'

Elsewhere the belief persists that animals are able to sense impending quakes. Earthworms and snakes appear above ground. Ants disappear. Cats and dogs 'act funny'. In the *very* short term, Hough mused, animals might react to a quake before humans because a lot of earthquakes have smaller earthquakes some minutes before, and the animals are more sensitive to these foreshocks. When she and her husband were studying in San Diego, they had a pet rabbit that lived in the house.

'There were a few times where she would thump on the floor and that's the first thing we would notice, and then we'd feel the shake.' For longer, more rigorous forecasting or prediction she's seen nothing scientifically convincing. The reporting is always retrospective. After a quake a cat owner thinks back and decides that their cat was acting strangely earlier in the day, forgetting all the other mornings when there was no quake but the cat still seemed to be acting strangely.

These days, prediction is something of a sore point for serious scientists. 'The USGS focuses its efforts on the long-term mitigation of earthquake hazards by helping to improve the safety of structures, rather than by trying to accomplish short-term predictions,' their web site admonishes.[28] It wasn't always so. Back in the '70s, people thought earthquake prediction was just around the corner. China, the Soviet Union and the USA all had prediction programmes. Chinese scientists claimed to have predicted the 7.5 Haicheng earthquake in 1975, saving many lives. American newspapers began publishing headlines such as 'Seismologists are getting close to a remarkable achievement: the ability to forecast with increasingly greater accuracy the location, time, and magnitude of earthquakes'.[29] They weren't. Possible earthquake precursors were tested and abandoned; there was talk of cherry-picked data; ideas foundered. In 1976 Chinese scientists failed to predict another magnitude 7.5 quake that struck the city of Tangshan, killing 250,000 according to the government estimate (many believe the true number to be two or three times higher).[30] Disillusioned, many scientists left the field.

'Even now if you go to some scientific meetings there are people talking about predictions,' Hough said, 'but where's the proof? If someone has a prediction method they should be able to start making predictions.' A part of me wonders whether among the attendees at these scientific meetings there's another Wegener – a man or woman with an idea that right now looks like 'delirious ravings' but that will come to be scientific orthodoxy. Perhaps – or perhaps these prediction methods are only the scientist's version of the non-scientist's yellowing sky, anxious dog. 'People say, here's a precursor we could have identified. This earthquake could have been predicted. But they're always looking back, and that's the problem,' Hough said. 'It's easier to fool yourself when you're looking retrospectively.'

*

Meanwhile in Salem, Oregon, a woman called Charlotte King wakes up with a headache and sharp heart pains, 'like small needle jabs'. She goes to update her web site: NEW TIMELINE SET FOR QUAKE 7.0 OR GREATER FEBRUARY 20 – FEBRUARY 28TH PLUS OR MINUS 12 HRS.[31]

King, who is in her seventies, describes herself as an earthquake-sensitive or 'biological forecaster'.[32] She believes that she is unusually sensitive to changes in the earth's electromagnetic field, which correspond to different areas of seismic activity, each one linked to a specific site on her body. In another post on her web site she writes that: 'If the quake is near Santa Monica you may experience some sense of a jolt, or vertigo with the quakes and with Palm Springs, Whittier, Pasadena and Burbank areas you can add, sharp heart pains with the sharp earaches both sides.'[33]

Like some sort of shaman figure or the priestess of an obscure animism cult, her writings suggest that she believes in a direct link between the remote, stately process of the plates, the tremors and shocks they send through the rocks, and the individual suffering human body. When she gives a forecast, she explicitly ties herself into the larger pattern not of human but of deep time.

Over the years – and especially before the widespread use of the internet – King was one of many people who wrote letters to the USGS predicting or forecasting earthquakes based, with varying degrees of scientific literacy, on such things as the shapes of clouds, barometric pressure, tides, magnetism, past earthquake patterns and lunar cycles. A woman called Linda Curtis used to date-stamp and file this correspondence, a collection of everything from scraps of paper to full-colour reports bound between plastic covers. These documents became known throughout the USGS as the X Files.

'There were a lot of individuals who were just lonely and looking to talk to somebody, and Linda was unusually patient with them,' Hough told me. Linda herself, in an interview with Ulin in *The Myth of Solid Ground*, said: 'Say someone predicted a seven in downtown LA, and we ignored it. Can you imagine the reaction if it actually happened? So this is sort of a little bit of insurance.'[34]

Charles Richter – inventor of the Richter scale (the more widely known precursor to the moment magnitude scale) – worked across the road at Caltech and received many such letters over the years. He said of the writers: 'A few such persons are mentally unbalanced, but most of them are sane – at least in the clinical or legal sense [...] What ails them is exaggerated ego plus imperfect or ineffective education, so that they have not absorbed one of the fundamental rules of science – self-criticism.'[35]

Ulin writes of Kathy Gori, whose predictions rely on headaches, and Zhong-hao Shou, who predicts earthquakes based on cloud formations allegedly created by heat released from within the shifting rock. Of John J. Joyce, who believes that electrical currents, not tectonic plates, are the key to understanding and predicting earthquakes, and of a man called Donald Dowdy – perhaps one of the not quite sane that Richter referred to – who believes that 'in the pattern of the LA freeway system, there is an apparition of a dove whose presence serves to restrain "the forces of the San Andreas Fault"'.[36]

The X Files correspondents attempt to rationalise, to find patterns, to bring the barely comprehensible processes of deep time into a comprehensible framework that domesticates the wild forces of the tectonic plates. 'We love to look for precursors in our lives, even day to day,' Scharer said to me. Or, as Stephen Jay Gould put it in *The Flamingo's Smile*: 'The human mind delights in finding pattern – so much so that we often mistake coincidence for profound meaning. No other habit of thought lies so deeply within the soul of a small creature trying to make sense of a complex world not constructed for it.'[37]

After our visit to Hollywood, Hough and I drove back to the USGS field office in Pasadena – a yellow two-storey colonial house that looks more like a family home than a cutting-edge research station. Inside, Hough introduced me to her colleague Stan Schwarz. I had wanted to see the X Files, but after Linda left, the collection was passed on to an outreach officer, who in turn left, and since then the whereabouts of the bulk of files are unknown. In the downstairs conference room of the field office, Schwarz, who looks after the computer systems that run the Southern California Seismic Network and has large flesh tunnels in both ears, showed me what was left. From the small sample I looked at, I concluded that women are more likely to experience earthquakes as physical manifestations in their bodies, men to dress up their research in the formal accoutrements of 'science'. That depressed me a little.

Schwartz passed over a sheaf of papers headed: 'Kenny Rogers of Corona, the only person in the world who has accurately predicted the timing, latitude and longitude of a large earthquake'. Based on charts made from plotting the patterns of previous earthquakes, Rogers started his predictions in the early '90s, after he retired from his job as an aeronautical engineer. He sent the USGS letters and elaborate booklets complete with maps, graphs and illustrations of ruptured highways, trucks sliding down embankments, one long slab of overpass broken off and crashed down on to the lower carriageway.

'Every two weeks he would send seventeen pages of predictions,' Schwarz said. 'We had several folders full; it went on for years.' Retired engineers, he theorised, are 'perfectly primed to turn into crackpots' because 'they know just enough science to think they really *know* science. But engineers are taught science as a bunch of facts, whereas scientists are taught it as a process.'

Turning the pages of the X Files, I imagined all the correspondents sitting in their houses compulsively working on charts and graphs and letters that no one except one or two USGS employees, one or two curious, gawping writers, would ever read. The X Files finally tell us more about the correspondents than about earthquakes. About their desperation to make sense of the world. Their need for the scientific community to acknowledge their work. Their desire to leave their names in the pages of the scientific literature, tied for ever with the great impersonal forces of the plates.

*

For the scientists, thinking about earthquakes and prediction is often a question of scale. In *Predicting the Unpredictable* Hough has written: 'On a geological time-scale they occur like clockwork. On a human time-scale they are vexingly, almost determinedly, irregular.'[38]

In part this is because we just don't have enough data. We haven't been recording information about earthquakes for long enough at a high enough resolution for a comprehendible pattern to emerge. 'With another 500 years of good data collection we could probably get over it, but for our lifetime there's an apparent randomness that I think we're stuck with,' Fryxell told me.

Hough, when I asked her if she thinks we'll ever be able to predict, said: 'Richter was asked whether prediction would ever be possible and his answer was, nothing is less predictable than the development of an active scientific field. I think that was very insightful. You never know what will be discovered. I've seen cases where seismologists said things with great authority that turned out to be wrong. We don't know everything. And some of the things we think we know we get wrong.'

Scharer shrugged when I put the same question to her. 'The thing is, even if we could predict earthquakes, you're not going to just stop LA for two weeks because an earthquake's coming. And the bridges are going to fall down whether we know it's happening on Tuesday or Friday ... If you haven't stored your water, then predicting isn't going to get you anywhere. If you haven't figured out how to slow down the trains when an earthquake's coming, you're still going to be in a lot of trouble ... I do find it useful to remind people of that.'

*

Back by the San Andreas, I got up and started walking along the fault, the pale mountains to my right. Eventually I came to a small, flat airstrip. A violently tanned man wearing a fluorescent green vest leaned against the back of a truck. He was part of a paragliding club. Today they were flying from Cloud Peak down to the airstrip. From up in the air, he said, you got a good view of the fault. We turned together and looked up towards the mountains. High in the blueness I saw the paragliders – tiny dark specks hanging beneath bright curves of colour.

Like the sort of painting that dissolves into a series of daubs and marks as you move towards it, you really need distance to get a handle on the San Andreas. Looking for the fault might be a metaphor for the search for deep time itself: so difficult to see what is happening from the ground or from inside human time, it's only when we step back or refocus our minds on that other scale that we catch a glimpse.

I stood still and watched as the paragliders arced in the air. From up in the sky the fault finally becomes visible. In places it is a long, puckering scar running through the landscape. Elsewhere, it shows up as lines of pale, churned earth like braided threads snaking along the base of the mountains. A thin ribbon of bright green trees. It is seen by the men and women hanging underneath their brightly coloured wings, flying from the North American plate across the line of San Andreas Fault and on to the Pacific.

7

A LOST OCEAN

On the British Geological Survey's map, chalk is represented by a swathe of pale, limey green that begins in Yorkshire, just below Scarborough, and curves in a sinuous green sweep down the east coast, breaking off where the Wash nibbles inland. In the south, the chalk centres on Salisbury Plain, radiating out in four great ridges: heading west, the Dorset Downs; heading east, the North Downs, the South Downs and the Chilterns.

The Chalk Escarpment is the single largest geological feature in Britain.[1] Stand on Oxford Street in the middle of London's West End and beneath you, beneath the made ground and the London Clay and the sands and gravels, is an immense block of white chalk lying there in the darkness like some vast subterranean iceberg, in places 200 m thick. Where I grew up in a suburb of Croydon at the edge of south London, this chalk rises up from underneath the clays and gravels to form the ridge of hills called the North Downs. These add drama to quiet streets of bungalows and inter-war semis: every so often a gap between the houses shows land falling away, sky opening up, the towers and lights of the city visible far in the distance.

The British Geological Survey (BGS) was established (under

a different name) in 1835, under the direction of the vice-president of the Geological Society of London, Henry De la Beche.[2] The world's first national geological survey, its original remit was to survey the nation and produce a series of geological maps. Today the BGS, which still produces the 'official' map of the UK's geology, is best described as a quasi-governmental organisation split between research, commercial projects and 'public good'. Quite a lot of its work is now done outside Britain: at the time of writing, ongoing projects include studies of groundwater in the Philippines and volcanic activity in the Afar region of Ethiopia. Since 1985 the headquarters have been at Keyworth in Nottinghamshire, where the BGS took over a former Catholic teacher-training college. The first year that the geologists were there overlapped with the last teacher-training cohort; two marriages resulted.

One week in early October four members of the BGS set up camp in a self-catering cottage near the town of Tring in the Chiltern Hills, about half-way between London and Oxford. They were on a training exercise as part of an ongoing project to produce a new geological map of the chalk of southern England. On the day I arrived, the wooden table in the main room was covered with maps, books, a half-drunk bottle of red wine and a packet of chocolate digestives. Field leader Andrew Farrant, tall and thin, with steel-rimmed glasses, was drinking a cup of tea. He had a sort of leather holster attached to his trousers from which swung a geological hammer with a surprisingly wicked-looking, long, pointed end.

As a schoolboy fossil-collector and caver growing up near the Cheddar Gorge, Farrant taught himself O-level geology and convinced his teachers to let him travel to another school to take the A-level. Studying geology at university, he first learned to map. Today most undergraduate geology courses in the UK still contain an element of training in the art of mapping – a fact I've heard some geologists, especially those in academia, deride. Mapping, they argue, is no longer cutting-edge geology, especially in the UK, where everything was essentially 'finished' over fifty years ago. It is the preserve of those gnarled old professors who can probably identify a rock type blindfolded but may struggle with the mathematics that lies behind so much world-leading work.

'I'd actually disagree with that position,' Farrant said.

Because as a student what the mapping does – even if it may seem a bit passé now, especially in areas that have been mapped a million times – is it forces you to think about all the different things and how they relate to one another – the bedrock, the superficial deposits [sands, gravels], the geomorphology, sedimentary environments, fossils, structure, the whole lot. If you end up working for a site investigation company or an oil and gas company, you need to be able to pull all that information together as a picture to find out what the site is like or where the oil is going to be. Field mapping teaches you to do that.

One of Farrant's colleagues said that mapping is a good way for a student to find out whether they're actually a decent geologist. You may get top marks in your exams, but can you apply that knowledge in the field?

Farrant has been working on the chalk-mapping project on and off since 1996. 'I would say that not enough attention is paid by the academic research community to understanding the geology of the UK,' he said. 'If I was doing this [mapping project] in east Greenland, then I'd probably get funding for it – east Greenland is sexy. And people tend to think that because we have a geological map of the UK it's all been done, but actually you can still improve it.'

The geology of the Chilterns, for example, was last mapped over a hundred years ago, in 1912. Since then the discipline has changed quite a bit. Geologists now know about plate tectonics and radiometric dating. There are lidar datasets (laser-based distance measurements) for elevation maps and digital terrain models and higher-definition Ordnance Survey maps, allowing hitherto unrecognised features to be recorded. All of this will affect the maps that are produced.

And when it comes to the chalk, these new maps matter in a way they didn't in 1912, because since 1912 the population of the south-east has increased by roughly a third.[3] In particular, this jump in the population has put pressure on the region's transport systems – often created by tunnelling though chalk to form such projects as HS2, the Gravesend tunnel and Crossrail – and the region's water resources – much of it stored in the chalk aquifer.

*

In 1746 the French geologist Jean-Étienne Guettard produced some of the first maps to show bands or zones of surface geological similarity.[4] Printed in black-and-white, and using dotted lines, shading and other symbols to represent the geology of France, they indicated a 'sandy zone', a 'marly zone' and a 'metalliferous zone'. Concerned with tracing rocks and minerals – and not with the time when the rocks and minerals appeared – this is properly a mineralogical rather than geological map. By the early 1800s, Georges Cuvier and Alexandre Brongniart in France, and William Smith in England, were working on what are often considered the first geological maps. The great leap was to show the rocks below the surface and to record their relative age and the manner of their deposition.[5] In 1810 Cuvier and Brongniart published a map of Paris and the surrounding areas. In 1815 Smith published the world's first true and comprehensive geological map of a country – England, Wales and (most of) Scotland.[6] In an age of gentlemen geologists Smith, a surveyor, was neither rich, nor posh, nor well-connected – in fact his social status barred him from membership of the august Geological Society of London – but he was obsessed with rocks, fossils and the idea of mapping the geology of Britain. He spent years travelling over the country to gather material, eventually bankrupting himself to produce the first copies of his map. He pioneered the use of fossils to identify rock types, writing in 1796 of 'that wonderful order and regularity with which nature has disposed of these singular productions [i.e. fossils], and assigned to each its peculiar stratum'. 'That is,' as an online Geological Society exhibition has it, 'he realised he could attribute specific fossils to particular strata, thereby allowing him to easily identify these formations across the [country].

Today one of the original copies hangs in the entrance hall of the Geological Society's headquarters in Piccadilly. When you pull back the blue velvet curtain protecting it from the light, one of first thing that strikes you is its beauty. The United Kingdom is furrowed by a series of curving lines running downwards right to left to reach a point around Taunton in Somerset. The country is a marbled mass of forest green, caramel brown, bubblegum pink, rich purple and pale lavender.

Looking at Smith's map, you can tell at a glance that the country is older in the west, younger in the east. That, roughly speaking, if you begin in the south-east and travel north-west up to the Highlands of Scotland, you travel back in time – from the newest formations of East Anglia to the ancient metamorphic rocks of the Highlands. Each stratum was given a different colour, based loosely on the colour of rock they indicated, and graded so that the strongest colour represents the base of the formation, lightening upwards. The colours Smith chose are, more or less, those still employed by all stratigraphers today. They are based on the colours of the rocks themselves: yellow for the Triassic sandstone of Shropshire, formed from hot, dry deserts; pale pink for Cambrian granites extruded from prehistoric volcanoes in what is now Wales; blue for the coal-bearing Carboniferous rocks of the Midlands, when that county was a land of seething, glistening swamps; pale, yellowish green for the white chalk, because white would have shown up badly against the paper.

Smith's map helped to shape the economic and scientific development of Britain during the Industrial Revolution. It showed where coal to power the factories might be found. Where clays and rocks to build the growing cities might be quarried. Where tin and lead and copper could be mined. Where a canal or railway line might most easily be dug. His map represented an increase not just in knowledge but also in wealth. But Smith, so the story goes, was ill-used by his contemporaries, including by members of the Geological Society, who borrowed his ideas without acknowledgement. Other maps were produced, making it difficult for Smith to recoup his own costs and leading him eventually to the King's Bench debtors' prison. It was not until 1831 that the Geological Society finally acknowledged his achievements, presenting him with the first Wollaston Medal, in 'recognition of his being a great and original discoverer in English Geology',[7] and not until 1832 that financial recompense finally arrived in the form of a government pension of £100 a year.[8]

Today Smith is sometimes known as 'the father of English

geology'. In 2003 one of his original maps was sold for £55,000. In Piccadilly, the Society that would once have refused him membership displays his relics like those of a saint: an oil painting complete with a lock of Smith's white hair sealed into the frame and two uncomfortable-looking wooden chairs.

*

The study of chalk is what is sometimes termed 'soft rock' geology. 'Soft rock' experts study sedimentary rocks such as sandstones and limestones, while their 'hard rock' counterparts work on the tough igneous and metamorphic rocks such as granites and slates. The categories aren't perfect – the calcite in sedimentary limestone, for example, is quite as hard as that of metamorphic marble – but the jargon sticks. Rivalry sometimes ensues. I once met a retired sedimentary geologist – now an enthusiastic Shakespearean actor in am-dram productions – who argued that 'soft rock men' are always the more thoughtful. It came, he mused, from thinking about the formation of sedimentary rocks. One rock unit formed from the quiet accretion of layers of sediment over many millions of years. The slow, slow formation of worlds. And what about hard rock geologists? I asked him. 'Hard rock men are all bastards,' he said.

The chalk world began to come into existence around 100 to 80 million years ago, when the Earth was entering a warming phase. Seas rose rapidly; one third of the landmasses present today disappeared beneath the rising waves. Geologists call this period the Cretaceous, after *Creta*, the Latin for 'chalk', and it is the longest geological time period on the stratigraphic chart: at 80 million years, it lasted far longer than the 65 million years that have elapsed since it ended. In regions where chalk is found today, the water was filled with many billions of microscopic organisms called coccoliths. When they died, their skeletons – disc-shaped calcite plates called coccospheres – sank down through the clear water, in such quantity that in places the ocean would have turned a milky blue. On the ocean floor the skeletons piled up, forming a soft, calcareous ooze. Over time this compacted and hardened. Living bones translated into white rock. The relative uniformity of the chalk, these massive thicknesses of rock, some a mile in depth, are testament to a stable, slowly drifting world where for millions of years nothing much happened.

During the late nineteenth century geologists began the task of further refining the existing rock units of type and time. The Jurassic-period Inferior Oolite, for example, has been divided into the Birdlip, Aston and Salperton formations, and each of these groups has again been subdivided. The Birdlip, though, covering a mere 2 million years, has itself been divided into seven sections. When it came to the chalk, however, the geologists divided it into three sections – Lower, Middle and Upper, each once covering 5 to 7 million years – and left it at that. There just wasn't very much to say about this regular, white rock, they felt, and furthermore there was little economic imperative to study the chalk in greater detail. There was some use of chalk as a fertiliser and, later, as a material to add to concrete, but it contains no coal, oil, precious minerals or metals, and is generally too soft to make a satisfactory building material.

Even among that subsection of the population who get excited by a good piece of rock, for many years the chalk was seen as fairly dull. When Farrant started work at the BGS in 1996, he told me: 'I got dumped on the chalk and I thought, Oh god how boring. When a colleague of mine got sent up to mid-Wales I thought he was lucky – much more interesting geology. It turns out I was wrong.'

In Britain – or more accurately the place that was to become Britain – the next big thing to happen to the chalk occurred around 50 million years ago, when the African plate crashed into Europe. The land buckled up, forming a series of ridges including the Pyrenees and the Alps. In Britain, a series of low chalk hills began to emerge from the sea. At first they were capped with mud and sandstones, but erosion eventually did its work and formed the bare chalk scarps of the South and North Downs and the Chilterns.

Today in the south-east of the UK, much of the chalk has disappeared underneath sprawling towns and inter-war suburbs, but where it hasn't been built over it produces a form of landscape often viewed as quintessentially English. Smooth hills – invariably described as 'rolling' – covered with short turf. Gentle slopes and steep escarpments, dry valleys and lonely beech hangers. Seen from a distance, a chalk landscape seems to ebb and swell like the ocean from which it once emerged. In December 1773, the vicar and naturalist Gilbert White, visiting friends in Ringmer, about ten miles from Eastbourne, wrote: 'For my own part, I think there is somewhat peculiarly sweet and amusing in the shapely figured aspect of chalk-hills in preference to those of stone, which are rugged, broken, abrupt, and shapeless.'[9]

On postcards and tea towels, images of chalk landscapes perform a particular version of Englishness, one that conjures up Vera Lynn, Shakespeare (who set the climax of *King Lear* on the top of the cliffs at Dover) and Rudyard Kipling (who wrote of the 'blunt, bow-headed, whale-backed Downs'[10]). 'Chalk has quite a central place in England's cultural history, the white cliffs of Dover and all that stuff, chalk downs and chalk streams,' Farrant said. 'And yet most people know nothing about what it is and how it formed.'

At the edge of the country the chalk becomes dramatic,

unsettling. Standing on the beach at Cuckmere Haven in Sussex, you look up at the towering whiteness and it seems for a moment as though it is falling towards you out of the blue sky. The exposed chalk has something cold and other-worldly about it. To see such whiteness, such brightness, feels unnatural.

From the south coast the chalk runs underneath the English Channel and reappears as another set of white cliffs, which the French call the Côte d'Albâtre ('Alabaster Coast') and the English tend not to talk about very much. These were much painted by Monet, Pissarro and Renoir. Chalk, which the English often seem to regard as peculiarly their own, lies under much of northern France, under bits of Scandinavia, under the Limburg province of the Netherlands and under parts of Germany, including the island of Rügen, whose white cliffs were painted by Caspar David Friedrich on his honeymoon in 1818 – a lovely dreamlike painting in which the cliffs look more like ice formations than rocks. A band of flint running through the chalk – known as the Seven Sisters flints – can be traced from the Yorkshire Wolds all the way to the Paris Basin. Other flint bands can be traced to Poland, as can seams of a clay known as marl.

In 1993 Richard Selley, then a professor at Imperial College London, had been thinking about the similarities between the chalk landscape of the North Downs and the Champagne region in north-east France.[11] His neighbour the engineer Adrian White had been unsuccessfully trying to farm sheep and pigs on his estate near Dorking in the North Downs. Why not try sparkling wine? Selley suggested. Today the vineyard – Denbies – is managed by White's son Chris. In 2018 they produced close to 1 million bottles of wine. Just over half were sparkling wines that would, if hailing from north-eastern France, be called champagne.

'Because of the similarities between the North and South Downs of southern England and the Champagne region of northern France, the south-east is known for producing very high-quality sparkling wines,' Chris White explained to me. 'We have the same temperature conditions, the *terroir* is the same, the south-facing slopes capture the maximum sunlight, the chalky soils make sure that the vines suck up the nutrients that they need and don't have their feet in water, which they hate.'

As Farrant said: 'The English Channel is really a minor thing. It's the same deposit basically, so there's no Brexit with the chalk.'

*

The Chiltern Hills run for 46 miles in a south-west to north-east diagonal from Goring-on-Thames in Oxfordshire to Hitchin in Hertfordshire. At their highest point – Haddington Hill, near Wendover in Buckinghamshire – a stone monument marks the 267-metre summit. Much of this landscape is farm-land. There are small villages huddled deep in the dry valleys, historic market towns and the edges of inter-war suburbia. I joined Farrant and his BGS colleagues there on a warm day of blue skies and strong, low autumn light. Many of the trees were still green, but some had begun to turn. Farrant and I set off with a new BGS recruit called Romaine Graham. Six months ago she had moved to the Survey from a job as a sedimentolo-gist with an oil and gas company. She'd been working on the chalk for two weeks and had blood blisters on the palms of her hands from wielding her hammer. A support bracelet protected one wrist.

We followed a track between hedgerows full of fat, red

rosehips and rambling old man's beard, with its white, fluffy seed-heads. We climbed over a barbed-wire fence between two ploughed fields; where there are no footpaths, the surveyors rely on the goodwill of landowners for access. Farmers are usually OK, but gamekeepers tend to be territorial. By the edge of the field Farrant and Graham used their hammers to break open pieces of chalk. 'This is the Zig Zag Chalk,' Farrant says. 'It's medium-hard, pale grey and blocky.'

We know now that the chalk was never just the three large, monolithic blocks of rock, and time, that the nineteenth-century geologists proposed – Lower, Middle and Upper. In the 1980s – around a hundred years after most of the other rocks were catalogued – geologists finally began subdividing the chalk into nine formations, each named after an area where a typical sample can be found.[12] The man behind much of this work is a geologist called Rory Mortimer. At a time when many geologists – the plate tectonics people, for instance – talk of the science as having moved from being 'descriptive' to 'quantitative', his classification work harks back to that of those nineteenth-century workers. (I once embarrassed myself on a field trip organised by the Geological Association by asking Mortimer what his favourite rock was. Hushed silence descended for a moment before the man next to me, in slightly scandalised tones, pointed out that I'd asked the rock preference of the country's foremost chalk expert. It was like asking a specialist in eighteenth-century paintings or sixteenth-century battle re-enactments which historical period they preferred.)

As we carried on walking, Farrant and Graham began to discuss differences between chalk formations. To the uninitiated, such differences can seem negligible. Working in chalk is all about getting your eye in, reading the subtlest of clues. The Zig Zag, for example, they described as 'rather dull,

John Major grey'. The Seaford, by contrast, is soft, smooth and bright white and often contains large flints. The Holywell is creamy white, filled with small fossils. The Lewes is white, creamy or yellowish. Chalk rock is very hard, closer to the hard limestones of Cheddar Gorge than the soft, crumbly white stuff that most of us think of as chalk: 'You wouldn't want to pick a fight with it.' Each formation represents a different world, and each of these worlds existed for far, far longer than humans have been on the planet.

Before Mortimer's work, engineers building on or through the chalk were always running into problems, Farrant said, because they treated, say, what used to be known as the Upper Chalk as one homogenous block. 'They didn't foresee that in some case they could be dealing variously with really hard chalk rock, or chalk with really massive flints, or really soft chalk with hardly any flints.'

Imagine stumbling, blindfolded, through an unknown landscape, uneven terrain underfoot, large, hard objects rearing out of nowhere. Without decent mapping, this is essentially the situation for a tunnelling engineer faced with an immense block of chalk. 'Obstructions are a very big issue,' Mike Black, Transport for London's principal geotechnical engineer, recalled in an interview in *New Civil Engineering*. 'We spend a huge amount of time on desk studies trying to work out where everything is or where it might be.'[13] An unexpected flint band or hard rock stratum can shatter the shield of a £100 million tunnel-boring machine. Hit a fracture or a seam of clay, and your tunnel – filled with men and machines – might flood with water. (Though tunnels had been built in chalk in the past, they were dug by hand, which allowed a greater margin for error. Because of the speed and power at which they move, the machines are, in fact, much more vulnerable than a human-wielded pick.) The

Channel Tunnel, for instance, doesn't go in a straight line from A to B but follows as much as possible a single stratum in the chalk that is one of the most suitable for tunnelling – the West Melbury Marly Chalk.[14] To plan the route, engineers looked at samples of chalk from boreholes and analysed the microfossils in order to map out the strata. 'That saved Eurotunnel probably half a billion pounds,' Farrant told me. Had they gone higher, they would have gone into the Zig Zag Chalk, which is more permeable and has more flints; had they gone lower, they would have hit a limestone bed, which is much harder and consequently difficult to tunnel through.

*

Where there are few outcrops, the surveyor must find other ways of getting at the chalk. Boreholes can provide information, but unless he or she can look at the borehole themselves, the surveyor is reliant on the quality and accuracy of someone else's interpretation of the data. Otherwise you look for old quarries and pits, badger setts, newly ploughed fields, even graveyards, where the earth has been recently turned. Working on a site at Stonehenge, Farrant found himself on his hands and knees looking for molehills beside the roar of the A303. 'Doing this work has got harder recently,' he said. 'Over the past ten years farmers stopped deep ploughing. Now they use something called a no-plough method, where they just put the seeds straight in the ground, which is fantastic for wildlife but for us it's a right pain.'

Up ahead he spotted a small copse, which he thought might contain the remains of an old chalk pit, dug, perhaps, by a farmer wanting lime to fertilise his soil, and dived into the undergrowth.

'We spend a lot of time fighting through bushes,' Graham said. 'Andy loves it.' By the time we caught up with him, he was sitting in the middle of the undergrowth hacking at a piece of chalk. 'Tottenhoe Stone,' he confirmed.

Graham leaned over and picked up a tubular flint, the shape of which she thought would amuse her boyfriend, also a geologist. 'We have fourteen boxes of rocks at home,' she said. 'I wanted to store them in the garage but he said that wasn't a safe enough environment for them. I mean,' she shook her head. 'I'm a geologist as well, but at the end of the day they're *rocks!*'

Where there are few outcrops, mapping the chalk also relies heavily on developing what Farrant calls 'landscape literacy': the ability to determine what is underground by studying the surface. Spring and autumn are the best times of year to do this sort of work – in summer crops and other vegetation obscure many landforms, and in winter there is the problem of snow, and the fact that it gets dark around 4 p.m. (In Wales, however, the geologists map in summer, to avoid the rain.) This is something that geologists have been doing since the time of Smith but, Farrant says, 'There's not much in the [academic] literature … It's knowledge that geologists have, but it's not really been written about formally.' Attaining landscape literacy might involve knowing that rounded hilltops are typically Seaford Chalk; flat fields typically Zig Zag. Or that where chalk is at the surface you find beech, yew and holly, and where the chalk is overlain with newer sands and gravels there are pine trees, heather and gorse.

By early afternoon the quality of the light had changed, become richer and more golden. The fields glowed lavender and apricot. Up close the soil was light grey and dry, and the surveyors' footprints looked like footprints on the moon. Standing at the base of Ivinghoe Beacon, the hill loomed above us, once the site of a Bronze Age barrow, and then an Iron Age fort, rising up abruptly from the farmland of the Vale of Aylesbury to form part of the ridge of the Chilterns.

'Doing this work, you're always trying to be in tune with the landscape,' Farrant said. Glancing at the hillside in front of us, he borrowed my notebook and deftly sketched out what he saw. 'Most people see it as a big slope with a hill on top and another slope going down the other side, but actually that steep slope has got various facets to it, so there'll be a slope, then a bench [a long, relatively narrow strip bounded by steeper slopes above and below] and then another slope … That's the Zig Zag Chalk, then the Holywell, then the New Pit and then the Seaford.' Then he pointed to a bump in the middle of the field in front of us. 'This, I would hypothesise, is the Tottenhoe Stone again. But we need some exposed chalk to check it.'

As we began to climb the Beacon, we passed an exposed bank of chalk, created when the path was cut into the hillside. Here the surveyors thought they might find fossils. Graham lent me her hammer. I'd never considered chalk a particularly fossiliferous rock, but after only five minutes of hacking we'd amassed a small collection of long-dead sea creatures. Pieces of chalk split in half to reveal a brown tubular worm, a brachiopod shell like a toenail, the perfect, concentric spiral of an ammonite.

'What I really love are trace fossils,' Graham said. 'They can tell you so much.' Trace fossils are the remains not of the creature itself but of its footprints, tracks, burrows, borings and faeces. 'Sometimes you can see two tracks or grazing trails – maybe two trilobites just skittering across the sand – and you can see where they join together for a bit, have a little party. It's easier to think about past landscapes when I can see traces of the creatures that lived in them. You think, wow, it was literally here.'

Building on techniques pioneered by Smith in the early nineteenth century, modern surveyors use fossils and microfossils to identify layers of chalk. Common bivalves (*Volviceramus, Platyceramus*) and echinoids (*Micraster*) in the lower part of the Seaford; common brachiopods (*Magas, Ctretirhynchia*) and oysters (*Pycnodonte*) in the upper Portsdown.[15] When the BGS store their fossils in Keyworth, they order them not according to species – as at the Natural History Museum – but according to stratigraphy, where they are found in the chalk layers.

Back in 2002, Farrant told me, police called on the help of local geologists when a tiny fragment of chalk was found underneath the wheel arch of the Soham murderer Ian Huntley. Two particular microfossils were discovered in the chalk: one found only in the Seaford and one only in the upper Newhaven. The presence of both microfossils meant that the chalk fragment could only have come from a specific two-metre thick layer – and the only place that chalk could have been driven over was a local farm track that a farmer had covered with that specific chalk, and where Huntley claimed he had never been. The chalk fragment formed part of the evidence that eventually secured his conviction.

*

At the top of the Beacon we sat down. It was very still and very silent. Somewhere up above a skylark was calling. From here you could see the fields of Buckinghamshire, Bedfordshire, Hertfordshire and Oxfordshire beyond. In the distance a row of small bushes flamed yellow and red, a line of fire along the edge of the green field. It made sense, I was thinking, that the first people to live here headed for this place, climbed up the hill to where a view of the world opened out.

Graham ate a banana and said that tomorrow she wanted to try and collect sloes. 'It's not always as pleasant as this,' Farrant warned me. 'You should come back when it's a freezing, raining day in January and we're stuck surveying some industrial estate in Watford.'

He got out a laptop and began to enter figures. The map they were working on is funded by the Environmental Agency, Thames Water and Affinity Water. Because chalk is highly permeable, water drains through it and the rock acts as a huge aquifer providing a source of drinking water. 'And the chalk here saves the UK economy billions because of not needing water treatment, but it's a complex beast.' The chalk acts as a natural filter, purifying the water that drains through it. But there are also fractures in the rock – and here the water flows instead of drains. Water companies need to know how the water flows through the chalk, where it can be safely extracted and how it can be protected from pollutants, such as nitrate run-off from farmers' fields. And to know how the water will flow through the chalk, to know what fracture patterns to expect, you again need an accurate, detailed map of the different formations. The Holywell fractures in a different way from the Seaford. A crack in the Newhaven is not the same as one in the Zig Zag.

When he'd finished with his laptop, Farrant pointed downhill. 'If you stood here during the Anglian Glaciation you would have seen an ice sheet coming right up to the base of the chalk scarp there.'

The next chapter of the story of the formation of the Chilterns took place around 450,000 years ago, when immense ice sheets covered the north of Britain, reaching down as far as

Watford. Beyond the ice, the Aga-loving, Barbour-wearing Chilterns was a wild expanse of cold, empty tundra. Unable to permeate the frozen ground, melting water from ice and snow and from summer rainfall flowed over the surface of the land, forming river channels that eventually cut down into the rock to create the dry valleys that are such a distinctive feature of the chalk landscape. 'The whole of southern England has been beautifully picked out by that periglacial weathering,' Farrant said. Further north, everything was just bulldozed by the ice. 'So the techniques we use down here for feature-mapping often don't work up north where you're just seeing the glacial over-print.' I pictured great blocks of ice moving remorselessly across a landscape. Ice like the blocks from Eliasson and Rosing's's *Ice Watch* but many, many times larger. Ice heavy enough to grind and smooth away the very rocks in its path.

Graham looked at her phone. It was time to go. The other surveyors would be finishing up and heading back to the cottage. Retracing our steps, the late afternoon light brought all the lines and angles of the hillside into focus, giving it that sculptural quality – a result of the periglacial weathering – that the painter Eric Ravilious captured in his inter-war landscapes of the South Downs. *Downs in Winter* (1934) turns bare, ploughed fields and curving hills into a series of simple, geometric forms beneath a pale winter sun. *The Lighthouse at Beachy Head* (1939) uses cross-hatching and green and brown washes to describe the cliff tops, while a white triangle becomes an exposed cliff face. *Chalk Paths* (1935) shows a line of black verticals – fence posts – cutting through the roll and swell of the land. 'I like definite shapes,' Ravilious, who grew up in nearby Eastbourne, once wrote.[16] He claimed he loved the South Downs because their 'design' was 'so beautifully obvious'.

*

A few weeks after my trip to the Chilterns I went for a walk on the North Downs, on the other side of London. Following a farm track towards the Ridgeway, the buzz and roar of the M25 was faint but insistent, like the distant rush of the ocean. Underfoot the path was pale brown and, where the thin topsoil had blown away, bright white – the bones of the land exposed.

Reaching the ridge, I paused, turned and saw London in the distance. Grey and silver towers coming up out of a muzzy blueness away over the beech trees and red-tiled suburban roofs.

As I stood there, looking back towards the city, it seemed as though the blueness intensified. And then it looked for a while as though the old Cretaceous ocean had returned to the London basin. Or as though I was seeing a flooded city some time in the future. I thought about melting ice sheets and sea-level rise and how, as I stood there, the south-east of the island was sinking while Scotland rose up – a see-saw effect caused when the great northern ice sheets began to melt around 20,000 years ago.[17]

And then in the city, I imagined, the ground would become heavy like a saturated sponge, the groundwater seeping up between the paving stones, bubbled up out of the drains and running along the gutters. The Thames would swell and over-top its banks. Fingers of brackish water creeping up Cheapside and into the grounds of St Paul's. The water rising over the Houses of Parliament, Big Ben, the Palace of Westminster. A blueness overtaking the landscape.

8

THE FIERY FIELDS

'We are going to the place where you can touch the volcano,' Vincenzo Morra told me.

In the heat of the afternoon we drove up into the low brown hills just west of Naples, past out-of-town discount stores and abandoned football pitches, and parked. Getting out of the car opposite a derelict, flat-roofed building, you noticed two things simultaneously: the smell – the rotten-egg stink of sulphur – and a great roaring, rushing sound, shockingly loud against the silence of the hillside.

Morra led me down a narrow track towards a small gully. The noise intensified. It sounded like an immense waterfall, a jet engine, some sort of infernal industrial process. And then at last we saw the cause: a great cloud of steam blasting upwards from a fissure in the grey rocks. Below, a pool of churning mud belched and bubbled. No vegetation grew near by. If you reached down and touched the ground, it was so hot that after a few moments you had to pull your hand back.

Campi Flegrei – the name means 'burning fields' – is what volcanologists call a caldera, meaning that it was formed when an older volcano erupted and collapsed in on itself, creating a gigantic bowl-shaped depression 12–15 km wide. Other examples include Yellowstone in Wyoming, Rabaul in Papua New Guinea and Sierra Negra in the Galapagos Islands. For hundreds of years this volcano has been dozing, announcing itself gently through hot springs and smoking fumaroles. Now, Morra has reason to believe that it is waking up. This is a big problem, anthropocentrically speaking, because Campi Flegrei is also home to more than half a million people, many living in the western suburbs of Naples and the town of Pozzuoli. It's as if we had learned that Leeds or Boston had been built on top of an active volcano.

'Everyone is scared of Vesuvius because they can see the cone, but actually Campi Flegrei is much more dangerous,' Morra told me, lighting a cigarette. 'With Vesuvius you know where the eruption will be. With Campi Flegrei, you don't know.' Vesuvius is what is called a composite volcano, so volcanologists know that the eruption is likely to come from the top or perhaps the side of the cone, but a caldera can erupt in many, many different locations.

'And it is not simple for 700,000 people to be evacuated. In an emergency people don't know what they need to do. If you asked my wife, she wouldn't know what to do. This is the problem.'

*

Just off the coast of southern Italy is the point where the African plate, travelling northwards, is closing up the Mediterranean Sea. One day you will be able to walk from Libya straight into Italy. Where the African plate is moving beneath the Eurasian plate, descending downwards into the mantle, some of the rocks melt, turning into magma. Sometimes that magma comes back to the surface via a volcano – a violent surface expression of the ceaseless churning of the sub-surface. It is these tectonic movements that are responsible for Campi Flegrei: A deep-time tectonic process exploding into human time.

To get a better sense of the size and shape of the volcano, Morra had promised to take me out on his boat, and at the harbour we met his friend Carmine Minopoli, a marine geologist, who would help with the sailing. 'Since Greek times – maybe earlier – people came to this area for the beauty and for the fertility of the soil,' Morra said as we headed out into the Bay of Pozzuoli, past a line of towering, sandy-coloured bluffs. It was peaceful on the water that day, but from the perspective of deep time the Campi Flegrei coastline is not calm but violent. Every bit of the landscape we were looking at had been formed by fierce volcanic activity. The sandy-coloured bluffs were made from compacted volcanic ashes. A smudge of bare rock in the distance was the remains of an old lava flow.

The generally accepted idea is that the Campi Flegrei caldera was formed around 35,000 years ago after something called the Campanian Ignimbrite super-eruption. Twenty thousand years later a second eruption tweaked the original caldera to give it its current shape. After that, there were many, many smaller eruptions within the caldera; I lost count of the number of times that Morra pointed out some hill or small island that was, in fact, the remains of a volcanic cone. Eventually the volcano fell silent, and during this lull Greek settlers arrived. They found rich soil, a pleasant climate and a wide blue bay. Later, several Roman emperors built summer retreats here, and the imperial fleet once found moorings at Pozzuoli. There would have been evidence of volcanism, as there is in much of the region – smoking fumaroles, hot springs, bright yellow sulphur deposits – but nothing obviously suggestive of the immense volcanic power hidden below the surface.

Considered on a human time-scale, the area around Campi Flegrei would have seemed like a good bet, generation after generation living more or less peacefully with their fishing boats and vineyards. Only someone able to look back at the history of the volcano in deep time – the history of its ancient eruptions – could have been prepared for the events of September 1538, when a volcanic vent opened in the ground near the spa village of Tripergole. The ensuing eruption was fairly modest but still buried the entire village underneath a cone of earth 133 m high and 700 m across. Eyewitness accounts describe the ground swelling and cracking and cold water gushing out, followed by great clouds of smoke and 'deep-coloured flames'.[1] Burning ashes and white-hot pumice were thrown 5.5 km into the air, and there was a 'noise like the discharge of a number of great artillery'. Birds fell dead from the sky until they covered the ground around the eruption site. Ash and pumice blanketed buildings and vegetation, forming a layer 25 cm thick in the town of Pozzuoli and 2–4 cm in Naples. The cone became known as Monte Nuovo, the new mountain. Over the next 400 years trees and houses appeared on its slopes. The volcano was quiet; the sense of danger receded. In a recent survey of residents, only 14 per cent of respondents mentioned Campi Flegrei when asked to list the active volcanoes in their area, and just 0.5 per cent of respondents listed volcanoes as one of the three greatest threats to their community. Unemployment and crime were more immediate concerns.[2]

Back on land, Morra gave me a lift to my hotel on the via Diocleziano, a long street filled with dust and petrol fumes that runs between tall apartment blocks, roofs bristling with aerials and satellite dishes. In the July heat, clothes dried on balconies, mopeds zipped through the morning traffic and locals sat outside the cafés that spilled onto the street. 'All of this is in the red zone,' Morra, said cheerfully. The red zone is the area of the city that will probably be destroyed when Campi Flegrei erupts.

Later that evening I ate at the Vulkania pizzeria – also in the red zone – where diners sit alongside a metre-high, cone-shaped model volcano. An unsubtle reminder to *carpe diem* and eat pizza? Tomorrow, after all, the sky may fall on our heads, and the smoke from a volcano blot out the bright Neapolitan sunlight. 'We can't fight nature,' was the opinion of a man sitting on a bench outside. He was seventy years old and his philosophy was: if it was going to happen, then it would happen. There was no point getting upset.

At night in my hotel in the red zone I watched YouTube videos of computer animations showing a potential Campi Flegrei eruption. Perhaps, I thought, you need to come to some internal accommodation with danger in order to live in a place like Campi Flegrei or Los Angeles or San Francisco. An acceptance that the slow and hidden processes of deep time may at any moment became fast and unhidden. On the screen a red and yellow plume formed, shooting upwards before falling down and spreading north, south, east and west, tongues of red and yellow rolling over the land and out into the sea.

*

At a moment in human history that sees us begin to grapple with the consequences of man-made climate change, volcanoes still feel grandly impervious to us, in the way that oceans, hurricanes and glaciers once used to. At the time of writing, the danger they present cannot be solved through engineering, or fully comprehended through science. They will be here long after we've been superseded by the rats or cockroaches or robots, and when a volcano erupts, the only thing a human can usefully do is get out of the way.

We have been trying to understand volcanoes for a long time. Perhaps the first volcanologist was Pliny the Younger – certainly he is credited as the author of the first written account of a volcanic eruption. Writing about the events at Vesuvius in AD 79, he describes 'a cloud [...] ascending, the appearance of which I cannot give you a more exact description of than by likening it to that of a pine tree, for it shot up to a great height in the form of a very tall trunk, which spread itself out at the top into a sort of branches'.[3] We now label that style of eruption 'Plinian'. Examples include the Mount St Helens eruption in Washington state in 1980 and Krakatoa, in Indonesia, in 1883.

The study of volcanoes divides into two main categories. First, there are modern volcanoes that are active and erupting on the Earth's surface today. These can be monitored and measured. Second, the rock record is examined to see where and how volcanoes erupted in the planet's deep time past. One afternoon in the UK, I went to University College London to meet Christopher Kilburn and find out more.

Kilburn was a student in Naples in the 1980s and still works closely with colleagues in that city. 'When you see an eruption for the first time you just want to do what everyone else does,' he told me. 'Take pictures and videos and what have you. Volcanoes are dramatic. I mean, I don't want to put palaeontology down but the straight-up truth is that it isn't quite the same thing ... Though perhaps we do have a few too many rugged people who like to be daring for no good purpose apart from showing off. There is an element of the "me-Tarzan" sort of thing.'

Volcanoes, he explained, are a bit like people. They have many shared general characteristics but come in different shapes and sizes, and each one has its own back-story, its own quirks that will determine when, and if, it erupts, and how large the eruption will be. In a general sense, though, volcanoes work as follows. The melting of solid rock to produce magma takes place at depths of between 50 and 200 km. Because it is lighter than the surrounding solid rock, the magma migrates upwards. In the case of Campi Flegrei, some magma appears to have travelled upwards from a large magma chamber at a depth of around 5 km to a point some 3 km beneath the surface. The big question is, what will happen next? As Kilburn explains:

It's not yet possible to make long-term eruption forecasts. Imagine that we've got to the stage where the magma has the possibility of escaping from where it's been stored. After that point the options are: one, it escapes but doesn't head towards the surface; two, it escapes, heads to the surface but doesn't have enough energy to break through the crust; three, it escapes to the surface and erupts. The challenge is knowing which of those things is going to happen, and at the moment we can't do that because, quite simply, we can't see through the rock to observe what the magma is doing.

Figuring out the behaviour of the underground magma is like attempting to piece together a complex jigsaw puzzle while blindfolded or, as another scientist I spoke to described it, like 'looking at a giant, fiendishly complex, plumbing system'.

There are many signals or indicators that suggest a volcano is nearing eruption. For volcanoes that have been quiet for long periods, the two signals you almost always see are changes to the surface of the volcano (ground deformation) and increased seismic activity (because of the movement of the magma.) At the vent Morra had showed me, and at various other places in Campi Flegrei, the volcanologists have rigged up an array of sensors and GPS devices, measuring the volcano's vital signals like anxious Lilliputian zookeepers with some giant, hibernating beast. Other signals that may be measured include increased vent temperature, changes to the vent geochemistry, increased CO_2 flux and a change in gas from hydrogen sulphide to sulphur dioxide. The last of these can be observed by the casual bystander: you know the switch has occurred when you stop smelling rotten eggs and start feeling your eyes sting.

'In the short term, as you get within days, you might be able to say that these signals are consistent with an approach to an eruption and sort of make forecasts approaching the quality of a weather forecast,' Kilburn explained. Ultimately, however, the presence of these signals doesn't guarantee that an eruption will occur. And while a weather forecaster may be forgiven the odd damp squib, when it comes to eruptions people tend to be somewhat more exacting. Diligent volcanologists cannot offer the certainty they demand.

But every time there is a major eruption, scientists learn something new. In 1995 Chances Peak on Montserrat erupted, and a year later Kilburn flew out to join a team monitoring the ongoing situation. In the Montserrat Volcano Observatory he happened upon a graph that showed a jagged upward curve of peaks and troughs, representing a series of earthquakes that had occurred prior to the eruption. He was reminded of a talk on common eruption trends given by the distinguished volcanologist Barry Voight. At the time of the talk Kilburn was working in a different field – lava flow – but Voight's words stayed with him. On Montserrat, looking at the graph, he could see a trend – the number of earthquakes had accelerated prior to the eruption.

During the '80s, volcanology – and geology more generally – was changing from an almost purely observational science to a more quantitative one, which sought mathematical patterns and built models. Before Voight, volcanologists made forecasts based on when phenomena witnessed before an eruption – such as earthquakes – reached a certain critical number. Voight's crucial insight lay in seeing that it was the rate at which physical processes changed that was important for making forecasts. Kilburn decided to use this insight to develop a model that could be applied to Campi Flegrei and used as a forecasting tool. To produce this model Kilburn needed to look at the underlying physics that determine when a rock fractures. 'It took bloody ages!' he said, shaking his head. Kilburn thought he'd be ready to publish by the millennium, but teaching, other projects, and a series of false starts intervened. It wasn't until May 2017 that his results finally appeared – and what they showed was extremely worrying.

*

The Vesuvius Observatory rents five floors of an unprepossessing glass-fronted office block in western Naples inside Campi Flegrei's red zone. 'So the Osservatorio Vesuviano, which has the duty of monitoring the eruption, will be evacuated before the occurrence of the eruption. This is nonsense of course!' Dr Francesca Bianco, the Observatory director, told me when we met. 'We are asking for a new site for our institute, but it is not simple.'

Founded in 1841 by King Ferdinand II of Bourbon, this is the world's oldest volcano observatory.[4] The original building – now a museum – is an elegant neo-classical structure on the slopes of Vesuvius. In the 1800s it attracted the sort of person who likes to run towards, instead of away from, rivers of molten rock spewing out of a mountainside. During the eruption of Vesuvius in 1872 a group of students 'driven by the curiosity to observe the phenomenon closely' were killed when a large flow of lava suddenly escaped from the north-west side of the volcano. A few days later the Observatory itself was dangerously surrounded by lava, but the director at the time, Luigi Palmieri, refused to leave, remaining in place in order to observe and record the phenomena. (Part of volcanology's darker glamour must come from the fact that its history books are littered with casualties. More recently the thirty-year-old American volcanologist David A. Johnston was killed during the Mount St Helens eruption in 1980. Moments before being overtaken by the pyroclastic flow at his monitoring post, he radioed his famous last words: 'Vancouver! Vancouver! This is it!'[5] His death was a continual source of guilt to fellow volcanologist Harry Glicken, who had swapped shifts with Johnston in order to attend an interview. In 1991 Glicken was himself killed in the eruption of Mount Unzen in Japan.)

Today, remote sensors at Campi Flegrei, Vesuvius and on the island of Ischia send information back to the main monitoring room at the Observatory, where most of the wall space is covered with banks of flat computer screens. On a desk in the centre of the room is a single red telephone: this is the emergency line that connects the Observatory to the Civil Protection Department in Rome. It's this phone that Bianco would use to contact the authorities if an eruption were forecast.

Bianco joined the Observatory thirty years ago as a student and, like all the other scientists I met in Naples, chose to stay and work in the area where she grew up. She is the Observatory's second female director and oversees around 100 members of staff, though the majority of other women are in the administrative department and only two are senior research-ers. 'Science is still in many ways a male environment,' she told me, 'but perhaps something is changing. For instance, the same day I became director here, the INVG [National Institute for Volcanology and Geophysics, the Observatory's parent body] also nominated their first female general director.'

It was back in 2005 that Bianco and her colleagues first noticed something worrying in the feeds from Campi Flegrei: the land in the caldera was moving upwards. This had happened three times before – in 1950–52, 1969–72 and 1982–84. On the last of these occasions the land rose by as much as two metres and the accompanying earthquakes forced around 40,000 residents to evacuate Pozzuoli – some never returned. Each time the land started moving, volcanologists waited nervously – seismicity plus ground deformation (in this instance uplift) being the two key indicators that an eruption is in the offing – but so far no eruption has occurred.

By December 2012, the new period of uplift, along with other signals including increased seismicity, convinced Morra and the other members of the Volcanic Risk Committee that the alert level for Campi Flegrei should be changed from green ('base') to yellow ('attention'). All monitoring activities at the Observatory were stepped up and new instruments brought in. Volcano hazard information packs were sent out to local schools. 'We give the information to the kids, and they pass it on to their parent and grandparents,' Bianco said.

In 2016 Giovanni Chiodini and other scientists from the INVG reported patterns of magma behaviour that resembled those seen before caldera eruptions at Sierra Negra and Rabaul.[6] 'The presence of more than half a million people living in the proximity of the caldera [...] highlights the urgency of obtaining a better understanding of [these] interactions,' they wrote.

A year later Kilburn published his model, which shed important new light on the rumblings.[7]

Eruptions occur when the Earth's crust stretches and breaks. Magma travels towards the surface, and the crust has to expand to accommodate it. Imagine an elastic band. You can stretch it so far, but there comes a point where the band snaps. Something analogous happens in the crust; when it breaks, an eruption may occur. Kilburn produced the first complete model of the relationship between the movement of the ground, the amount of seismicity and the chance of the crust breaking. '[The model] brings a structure that allows us to quantify all these things,' he said. 'Up until now it's all been gut feeling rather than actual objective methods.'

Applied to Campi Flegrei, Kilburn's model challenged the prevailing assumption that during the periods of unrest between 1950 and 1985 the stretched crust had returned each time to something like its original state. In fact, the apparently separate periods were all part of one prolonged sequence of seismic activity. If Kilburn is correct, the next major uplift is not going to start from where the last one began; it's going to start from where the last one finished. This means that the volcano's crust is getting ever closer to breaking point. 'And the difficulty is that people might have been frightened in the first incident but they're less frightened in the second and even less in the third ... And of course it should be exactly the other way around! You don't want to frighten people unnecessarily, but you want to get the message across,' he said.

*

The Civil Protection Department (CPD) envisages four possible scenarios for a Campi Flegrei eruption: an explosive eruption, which will be classed as small, medium, large or very large; multiple, simultaneous eruptions from different vents; a phreatic eruption, which is driven by steam; and an effusive eruption, in which the lava flows steadily. They have calculated that there is a 95 per cent chance that the eruption will be no bigger than medium-size. Because of the large number of people in the caldera, this will still be incredibly dangerous. If you're in the wrong place at the wrong time, a very small eruption will kill you as effectively as a large one.

But more spectacular conflagrations have happened before and may happen again. The Agnano Monte-Spina eruption, for example, took place in the caldera 4,100 years ago.[8] Some volcanologists use it as a point of reference for a 'large' Campi Flegrei eruption. In such a scenario, repeated earthquakes would shake the ground. A large grey cloud would spread over the caldera, swathing it in darkness. This cloud would be formed from a column of hot water, toxic gases, ash and pumice propelled up to 25 km into the air. Anyone caught near the eruption vent would be bombarded with fragments of white-hot rocks and boulders. Deposits of hot ash, thick enough to destroy buildings, would fall as far as 40 km away – the distance from London to Guildford, or Edinburgh to Falkirk, or New York to Edison in New Jersey.

'Ash fall is particularly nasty,' Dr Amy Donovan, a lecturer in geohazards at the University of Cambridge, said when we spoke on Skype. 'I've been at volcanoes where it just goes completely dark and you have to get somebody else to walk in front of the truck to show you where the road is. Driving in really thick ash fall messes with the cars as well. It's like flour but it's basically rock, so it doesn't just brush off, and it gets into anything mechanical and into electronics. When there's a lot of it you can't really breathe outside without a mask.'

Even nastier is what happens when the eruption column collapses – the pyroclastic flow. Advice on the CPD's web site is brief and to the point: *The only defence ... is moving away in advance from the area that could be hit by this eruptive phenomenon.* A pyroclastic flow, Morra explained, is far more dangerous than lava. Lava is typically slow-moving (there are exceptions). You can run and even walk away from it. In Iceland, communities have managed to guide lava flows away from their towns and harbours. No one can guide a pyroclastic flow. Reaching speeds greater than 60 m.p.h., they are lethally hot currents of hot gases and rock particles (tephra) that reach between 200 and 700°C. Those that contain less rock and more gas are known as pyroclastic surges. YouTube videos of pyroclastic flows show immense roiling brown clouds that cover everything in their path. All trees and buildings are flattened. Layers of debris pile up hundreds of metres thick. In such conditions the extreme heat will kill any living organism in a fraction of a second. The pyroclastic flows at Pompeii produced temperatures of up to 300°C, which would have caused the victims to spasm into contorted poses, like those found among the famous plaster casts.

For Bianco and the Observatory staff, one of the greatest challenges will be deciding when to trigger the final red alert. There are currently no set criteria to follow. (Kilburn's model may explain crust failure, but even that does not guarantee an eruption.) 'A lot still involves considerable amounts of expert judgement. What have you seen before? How have you managed before?' Donovan explained. Because major eruptions are relatively rare, it can take a lifetime to build up that expert knowledge. The USGS, for example, is currently facing the retirement of a tranche of experienced volcanologists who have been through a number of major eruptions and must consider how best to preserve their expertise.

The stakes are incredibly high. In the L'Aquila earthquake in Italy in 2009 (a low-probability event with high stakes, much like an eruption), more than 300 people died. In 2012 six scientists and a government official were found guilty of manslaughter and sentenced to six years in prison. Prosecutors argued that the individuals – all members of the government risk-assessment commission – failed to communicate properly the increased risk of a major earthquake in the wake of smaller tremors that preceded the quake. Some bereaved families claimed that it was because of reassuring official statements that their relatives made the fatal decision to stay indoors when the quake struck. Critics countered that there is no precise method for predicting an earthquake, and that the trial would serve as a disincentive for scientists to advise the government on future occasions. The convictions of the scientists were overturned on appeal in 2014.

At the other extreme, volcanology is still haunted by the example of the 1976 Guadeloupe eruptive crisis, when 72,000 people were evacuated for between three and nine months, at huge economic and personal cost. The volcano never erupted.

'And it can be very hard to get a government to prioritise volcanic risk because government terms are very short compared to volcano terms, which can last for hundreds of years, even tens of thousands of years,' Donovan said. It can be difficult, psychologically and practically, to plan for a deep-time event: easier to hope that it won't occur during your relatively brief watch. (Globally, this problem is compounded by the fact that volcanic risk is highest in the developing world, where many governments have a plethora of other more obviously immediate concerns to worry about and devote their resources to.)

When the Campi Flegrei red alert is triggered, the heads of the emergency services and the scientific and technical advisers will meet at the CPD's headquarters in Rome. One morning in Naples I left my hotel to catch the train north. Wildfires were burning on the hillsides around the city, and the air was hot and hazy and filled with the scent of bonfires. Rome, if anything, was hotter, and it was a relief to arrive at the CPD's air-conditioned offices. Here a belt-and-braces approach to safety is observed: plenty of gleaming modern technology but also a crucifix on the wall and, in the small vestibule in front of the Operational Committee Room, two chairs positioned in front of a richly painted gold icon.

'We have calculated that seventy-two hours is the minimum amount of time we need to complete the evacuation,' David Fabi from the emergency management office told me. This breaks down as twelve hours for organisation, forty-eight hours for moving and an extra twelve-hour security margin. It will require a mammoth feat of organisation. Fabi and his colleagues wear blue polo shirts and look action-ready, as though they might at any moment jump into a helicopter. They have divided the red zone into twelve sectors: six municipalities and six Neapolitan neighbourhoods. When I visited, they were updating the evacuation plans and had sent out questionnaires asking residents to think about how, in an emergency, they would leave the red zone. It is assumed that many people will drive their own vehicles – though there are concerns about the quality of the local, rural roads and whether or not they can support such a large amount of traffic. During the evacuation, radio, internet and television will be used to communicate with residents, and additional methods of transport will be provided including around 500 buses, 220 trains and a number of ships for each day of the evacuation. Each of the twelve sectors has been 'twinned' with another area of Italy, where the evacuees will be hosted. For example, residents from the municipality of

Pozzuoli will travel to Lombardy, and residents of the Naples neighbourhood of Chiaia will go to Sicily.

All of this is complicated by the fact that no one knows exactly where the eruption will occur, and which access routes out of the caldera might be blocked off. And it assumes that there will be those crucial seventy-two hours of warning. In his report on the 1994 eruption at Rabaul, Hugh Davies, a professor of geology at the University of Papua New Guinea, wrote that there was now 'clear evidence that a caldera-collapse volcano can erupt with as little as 27 hours of precursor activity [... this] has implications for those charged with the monitoring of other caldera volcanoes that are close to population centres, notably Long Valley caldera in California and Campi Flegrei near Naples'.[9] Indeed, the speed with which the Rabaul eruption developed meant that the authorities effectively had only twelve hours' warning, making it impossible to put into place many of the pre-eruption safety measures, such as the distribution of food to care centres, the orderly evacuation of Nonga Hospital and the relocation of the disaster headquarters and radio station.

Communication between the scientists, the authorities and the population will be the key to a successful evacuation at Campi Flegrei. At Rabaul, despite having only twelve hours' warning to evacuate 45,000 people, only five residents lost their lives (four killed by direct volcanic effects and one by lightning). In his report Davies speculates that the high level of hazard-awareness among the community was a major factor in the evacuation's success. In Campi Flegrei, Kilburn has been interviewing residents who witnessed the evacuations at Pozzuoli in the 1970s and '80s, and the results, he explained via email, are alarming. There is 'an underlying cynicism about whether evacuations are motivated wholly by scientific concern'. A recurring theme is that evacuations benefit property speculators, who have taken advantage of the collapse in housing prices in Pozzuoli during emergencies. 'It is not clear that the suspicions are well founded. However, even if they are incorrect, if the allegations are *believed* to be true, they encourage resistance to accepting the warnings that volcanic activity is possible.'

Donovan agreed. 'Trust is hard won and easily lost. Some of it is about communication – it's about getting people to understand how much uncertainty there is in forecasting and that the scientists are not deliberately withholding or giving bad information – they just don't know exactly what is going to happen and when.'

This distrust of scientists and government authorities is one of many factors that may make people reluctant to leave their homes in the hours before the eruption takes place. Other concerns include fear of looting and worries about pets. 'Can you take your cat with you, or do you have to leave it to the mercy of the volcano? That can be a big issue for a lot of people, especially in developed countries,' Donovan said. For others, the prospect of living in temporary shelters is a major deterrent. During the evacuation of Montserrat in 1995 several elderly people chose to remain – and die – in their homes rather than face the unsanitary and undignified conditions in temporary, overcrowded evacuation shelters. And added to all this is the uncertainty about when – if ever – the evacuees will be able to return home: a volcanic eruption can last for months or even years.

'We have to try harder to listen to the population, what they don't know, what they want to know, and how that information can be made accessible,' Kilburn wrote. 'The key message is that the effective delivery of a forecast is as important at getting the forecast right in the first place.'

＊

In the rocky gully west of Naples, I turned away from the vent to look down the browning hillside to where the houses and shops and apartment blocks of the red zone shimmered in the heat, and where, on the fertile volcanic soil, pink and white and purple bougainvillea grew in startlingly bright profusion along the roadside. Further away were strips of vineyards, and yellow and terracotta buildings clustered along the shoreline of the bay.

Considered from the perspective of deep time, everything I could see had come about because of the volcano. Thirty-five thousand years earlier it had collapsed in on itself and formed the bowl of the caldera. Falling ash, compacted, had become rock; lava flows had solidified into outcrops; old volcanic cones had become tree-covered hills. Later, the rich volcanic soil encouraged people to settle inside the caldera, and their descendants took the rocks formed from volcanic ash and built the houses and apartment blocks that wind up the hillside.

What this landscape will look like in the future depends very much on what the volcano does next. Three kilometres beneath the hills and streets and fields of Campi Flegrei, the magma is stirring in its subterranean chamber. Up on the surface, the scientists bend anxiously over their instruments and computers and watch the jagged lines that run, left to right, across their screens. Government staff work on a complex series of plans that they hope will never be put into action. The residents of the red zone go about their daily lives, measured out in human time, in days and years and decades, births and deaths, job interviews and holidays, roads taken and not taken. Sometimes a journalist arrives – often foreign – and asks them, *But what is it like? Are you scared? Are you worried?* The residents of the red zone shrug. They have always lived here. Their parents and their grandparents and great-grandparents have always lived here among the hot springs and sulphurous vents.

I turned back to look again at the steam and vapour escaping from the fissure, prodded, with the toe of my boot, a charred stub of vegetation. Morra had walked closer to the vent. 'For me, this is something very beautiful,' he said, smiling. 'The breath of the volcano.'

PLANTS AND CREATURES

9

AMMONITE

The man who owned the fossil shop at Bell Cliff had moved out for a month because of the filming, he said. Now he was looking after the fossil shop on Bridge Street while that proprietor – his son – took a holiday. The fossil shop on Bridge Street is opposite Lyme Regis Museum, which is built on the site of what was, in the early 1800s, one of Lyme's first fossil or 'curie' shops, presided over by Mary Anning: a poorly educated, working-class woman who has been called the 'unsung hero of fossil discovery'[1] and 'the greatest fossilist the world has ever known'.[2] *Ammonite*, the film in production when I arrived in town, is based on her life.

Lyme Regis is a small town on Dorset's Jurassic Coast. During the week, crowds had gathered there to catch a glimpse of the film's stars, Kate Winslet, Saoirse Ronan and Fiona Shaw. Winslet had bought a coat from a local shop and was seen drinking a pint in one of the town's public houses. Coombe Street, where I was staying, would be closed to traffic and strewn with nineteenth-century mud to film a carriage scene. Around Bell Cliff, the old walls of the promenade had been covered with more authentically old-looking polystyrene walls. A man in a hi-vis jacket stood guard over a collection of props: wooden boxes and barrels and coils of rope. The Bell Cliff fossil shop had a new sign: Anning Fossils. Everything inside had been ripped out – including the electrics. The owner, a retired builder turned fossil trader, was considering keeping the Anning sign above the door. It would be good for business.

'But what I don't understand,' he said, looking suddenly awkward, 'is why they have to go into all that other stuff.'

Reading the press when *Ammonite* was first announced, you could have been forgiven for thinking that the most inter-esting thing about Mary Anning was that around two centuries after her death – about the time it takes for 7 mm of the local Blue Lias rock formation to accrete – Kate Winslet and Saoirse Ronan would star as gay lovers in her biopic.

About this situation Lyme was undecided. Some residents expressed disapproval. 'There's no evidence for her having had a lesbian relationship,' several said to me. The *Telegraph* quoted an Anning descendant called Barbara: 'I believe if Mary Anning was gay she should be portrayed as gay … But I do not believe there is any evidence portraying her as a gay woman.'[3]

Others were enthusiastic. 'For too long this woman has been unknown and lost to many. We see only good coming from this film,' tweeted Mary Anning Rocks, a group started by an eleven-year-old Lyme resident called Evie to campaign for the commission of a Mary Anning statue. 'The newspapers have been making a lot of song and dance about it, but I don't really see what the problem is,' a woman in a café on Broad Street said. 'And you can't just have lady finds fossil, end of story – that would be a pretty boring film.' Her friend nodded: 'No one lives their life like that anyway,' she said. 'There's always some kind of love interest.'

Finally, an annoyed-sounding Lee tweeted that: (i) none of the journalists or Lyme residents had seen the film yet, given that it hadn't been made, and that it might be better to watch it before critiquing; and (ii) since there was no evidence of any heterosexual relationship either, why assume she was straight?

*

In 2019 visitors to Lyme Regis could choose between the Mary Anning tour (£8), the Jane Austen tour (£10) and the John Fowles's *The French Lieutenant's Woman* tour (£75 for a group of three).

Fowles was curator of the Lyme Regis Museum for ten years – where he did much to promote the work of Mary Anning – and the 1981 film of his novel, starring Meryl Streep, was shot in the town, bringing a wave of tourists that Lyme hopes *Ammonite* will replicate. Austen visited Lyme in 1803 and 1804, and later made it the setting for part of what I think her best novel, *Persuasion*. (Today a gift shop called Persuasion, selling 'coastal inspired gifts and clothing', can be found near the old stone breakwater.) In 1804 Austen asked the cabinetmaker Richard Anning – Mary's father – to mend a piece of furniture.[4] Whether she encountered Anning's daughter is not recorded, but certainly, if she had, it is difficult to imagine that anyone could have guessed that this twenty-nine-year-old woman and five-year-old girl would go on to become part of the fabric of the town. Austen didn't, in the end, take up Mr Anning's services – far too expensive, she wrote in a letter to a friend.

The tour guide Natalie Manifold used to come to Lyme as a child because her mother enjoyed fossil-hunting and *The French Lieutenant's Woman*. After university she moved to the town. 'When I first started – around ten years ago – Jane Austen was the most popular tour,' Manifold said. 'Now it's Mary

Anning.' When occasion demands – at the Lyme Regis Fossil Festival, for example – she dresses up as Anning, donning a custom-made version of the mossy green coat and red-ribboned straw bonnet that Anning wears in the only portrait completed during her lifetime. 'She wouldn't have been wearing that kind of thing to go fossil-hunting, though,' Manifold said, showing me a pen-and-ink sketch of Anning on the beach wearing a voluminous checked skirt and a top hat – an attempt to protect herself from falling rocks.

From our twenty-first-century vantage point, much of Anning's nineteenth-century life is inaccessible, buried from sight, but glimpses appear. A handful of moments have been preserved in contemporary accounts, standing out in the historical record like a fossil jutting out of the cliffs.

'During a Lyme horse show a storm developed, and after a terrific flash of lightning three women and a baby were seen lying on the ground under an elm tree,' Manifold told me. The women were dead, but the baby, placed in a bath of warm water, revived. The baby was Anning. A contemporary account records that she 'had been a dull child before, but after this accident became lively and intelligent and grew up so'. Later, Anning's father taught her and her brother Joseph to hunt for fossils to sell to passing visitors and local collectors: ammonites, belemnites, devil's toenails, fossil fish. While we now scour the pages of history for evidence of Mary's life, she was learning how to read what had been recorded in the larger, stranger book of deep time.

When Richard Anning died of consumption, the siblings carried on the trade to supplement the family income, and then, when Mary was twelve years old, her brother found something very unusual embedded in the cliffs.[5]

The fossil skull looked a little like a crocodile – but when had crocodiles had quite such beak-like, pointed snouts and such strangely plated, circular eyes? Mary spent a year searching for the body and then painstakingly digging out the outline of its 5.2-metre-long skeleton.

People in Lyme still called it a crocodile. Science eventually named it *Ichthyosaurus*, or 'fish lizard'. We know now that it was neither fish nor lizard, but a marine reptile that lived 201–194 million years ago, when the place we call Lyme Regis would have been at the bottom of a Jurassic ocean. The 'crocodile' was eventually sold for £23 to a keen local fossil collector, and later to the British Museum for £47 5s.[6] It survives today in the Natural History Museum in London. Not yet a teenager, Mary had made her first major discovery.

*

The long grey arm of the Cobb curves around the harbour. It was here, on the stone breakwater, that Austen's Louisa Musgrove fell and injured herself, and Fowles's Sarah Woodruff stood staring enigmatically out to sea. Away from the shore the white and pink painted buildings of the old town – a pretty jumble of narrow lanes – climb uphill, as though jostling to escape the water.

Because of the sea, and because of the layering of the rocks – soft, sliding muds sandwiched between heavy limestones – the cliffs surrounding Lyme are always shifting and stirring. To the east of the town is a place called Black Ven, one of the largest mudslides in Europe. Several years ago the Department for Environment, Food and Rural Affairs (Defra) and the local council spent £19.5 million on a new sea wall and other defences – but standing on a narrow strip of beach at high tide, staring up at the soft, grey-blue cliffs, you have the feeling that anything man does here can only be a very temporary staying.[7] If land wants to move then, eventually, that is what it will do. The cliffs slump downwards in a cascading set of levels – like one of those penny-pusher seaside arcade games with the ledges of coins that, every so often, overspill. When this happens, tonnes of rock and mud fall on to the beach below, leaving behind a thick, black swipe like an oily hand wiped across the cliff face.

But it is this constant churning that brings the fossils to the surface – and after every storm, every landslip, the hunters or fossilists, as they are known, will be out turning over the newly broken ground. There had been no storm the morning I went down to the beach, but it was so early that the sun, sea and sky were pale, leached of colour, the sea white at the horizon line, and the steps leading down to the beach were still underwater. And so for a while I had the place to myself, walking along the narrow strip of damp rocks and pebbles between the muddy cliff and the slop and crash of the breaking waves and the sucking sound of the water running back over pebbles.

'We're doing what the eighteenth-century geologists did,' the palaeontologist Jan Zalasiewicz had said to me when we went looking for ammonites in the railway cutting in Leicester. Methods of collecting haven't really changed much in the last 200 years. There are power tools now to break apart the larger rocks, but fossilists and bone-hunters still go out with a hammer and chisel. They still use plaster of Paris to blanket a bone prior to removing it. As I walked on the beach, I thought about Anning making her way across the foreshore in top hat and checked skirts.

The soft, dark cliffs at Lyme are part of a formation known as the Blue Lias, formed during the Jurassic. Rib-like layers of pale grey limestone alternate with wodges of thick grey-blue shale; as a whole, the cliff looks like a grainy ultrasound image. The different layers show that this Jurassic ocean was by turns dull and muddy (the shale), and clear, warm and shallow, filled with corals and sea lilies and the first scuttling crabs (the limestone). Because of a recent landslip, a Victorian rubbish dump that was once on the cliff top is now half-way down the cliff and on the beach. Every so often among the boulders you find a rusting object – an old fork, a piece of railing. These man-made objects weather out of the rocks just as the fossils do. Pieces of the past become the present.

After a while, other solitaries began to appear from the direction of Charmouth. They had the slow, stooped gaze of the fossil-hunter. Watching from the distance you would think they were drunk or ill, making slow progress as they weaved from side to side, turning round and round over the same patch of beach.

'Found anything?' we called to one another. A man wearing black Crocs and a tattered black hat opened up his fist to show a handful of slender belemnites – bullet-shaped fossils of extinct cephalopods. Another man was hunting for coprolites – fossilised faeces. But Lyme, he said, was over-collected these days. It was getting more difficult to find anything good. He blamed the tourist board and social media.

*

To be a female fossilist in the early 1800s was to be peculiar, strange, unfeminine. One contemporary account describes Anning as 'masculine in expression'.[8] Another talks disparagingly of finding her 'in a little dirty shop, with hundreds of specimens piled around her in the greatest disorder [...] a prim, pedantic vinegar looking, thin female; shrewd and rather satirical in her conversation'.[9] At Lyme, with its tides and high, unstable cliffs, it was also dangerous. In 1833 a large rockfall missed Anning by inches and killed her dog Tray. Anning persevered, driven, we might imagine, by a mixture of financial need and scientific curiosity, clambering over the treacherous rocks to extract with difficulty her reptile bones, at once brutally heavy and terrifyingly fragile.

In addition to the ichthyosaurs, Anning's major finds included the first complete *Plesiosaurus*, a *Squaloraja* fish skeleton that proved the missing link between sharks and fish, and the first British example of the flying reptiles known as pterosaurs. She pioneered the study of coprolites – a new and valuable source of information for palaeontologists. Her reputation grew as members of the Geological Society and other fossil enthusiasts visited her in Lyme to go fossiling or buy specimens. Charles Dickens noted the esteem in which she came to be held by the learned men of the day, including Richard Owen (who coined the word *Dinosauria*), William Buckland (who wrote the first full account of a fossil dinosaur), Henry De la Beche (first director of the Geological Survey of Great Britain and, despite the difference in social class, a childhood friend of Anning) and the French palaeontologist Georges Cuvier.[10]

Around the same time another visitor to Lyme, a Lady Silvester, wrote in her diary that: 'It is certainly a wonderful instance of divine favour that this poor, ignorant girl should be so blessed, for by reading and application she has arrived to that degree of knowledge as to be in the habit of writing and talking with professors and other clever men on the subject, and they all acknowledge that she understands more of the science than anyone else in this kingdom.'[11] More than just a hunter of fossils, she had, Zalasiewicz has written, 'a keen forensic curiosity about the creatures themselves and an undisputed brilliance in practical fossil anatomy'.[12]

Despite this, Anning's contributions to the new science were not recorded in the scientific papers written by the men who came to learn from her, and she would publish nothing herself except one letter in an academic journal disputing the genus of a shark fossil. Her work as a hunter and gatherer, as was tradition at the time, was not recorded in the catalogues of the museums that displayed her finds.

'Miss Anning's business, of course, was not to take sides, but to furnish the combatants with munitions of war – now a paddle, then a jaw, then a stomach full of half digested fish', Dickens wrote, but there is evidence that Anning resented this position.[13] A young woman named Anna Piney, who sometimes joined Anning on her digs, wrote of her friend's frustrations: 'She says the world has used her ill [...] these men of learning have sucked her brains, and made a great deal of publishing works, of which she furnished the contents, while she derived none of the advantages.'[14] Anning would probably be happy to learn that today her name is better known than many of those male contemporaries.

It is tempting to wonder also what she would make of the position of women in in science today. Around 40 per cent of entrants to geoscience degrees in the UK and the US are female[15] but these numbers don't translate up the academic ladder. The analogy often used is that of a pipeline towards professorship from which women disproportionately 'leak' out. In the UK, women make up just 17 per cent of Earth, marine and environmental science professors,[16] while in the US the figure is 15 per cent.[17] Outside of academia, women account for only around 24 per cent of the core-STEM (Science, Technology, Engineering and Mathematics) workforce in the UK[18] and 28 per cent of the science and engineering workforce in the US.[19]

'I think that the challenges today for women are in the area of unconscious bias,' the UCLA geodynamicist Carolina Lithgow-Bertelloni told me. 'Of people not giving you your due, not thinking of you. Given the same level of achievement as a male colleague, women don't get the same chances – there's no doubt about that. At first people won't listen to you but they'll listen to the guy saying the same thing.'

When Maria McNamara, a palaeobiologist now at the University College Cork, was pregnant, a (female) colleague warned her against taking maternity leave: it would ruin her career. She took the leave anyway and was still breastfeeding when she returned to work, expressing milk twice a day. 'You're doing what you're told is the best thing to do for a baby, but people will shoot you a glance,' she said. 'They'll be making a note if you have to leave a meeting.' Last year McNamara spearheaded an initiative by the Palaeontological Society to provide grants for breastfeeding women to bring their partners to international conferences, allowing the women to travel.

'When I look at successful women in science, very often they are painted negatively,' McNamara said. 'Oh she's career-driven. She's really *determined*. You'd never say that about a guy.'

'You are aware of it, you are outraged by it, but it cannot drive you,' Lithgow-Bertelloni said. 'Because then you think of nothing else and it interferes with your science. When you're forming an equation or coding, it doesn't care if you're a man or a woman. When you're writing a paper, it doesn't care.'

*

After something dies its body disintegrates, is broken up by other creatures or by the processes of decay. But in very rare instances fossilisation occurs and the material of the body

– typically the hard parts, such as shell or bone – is replaced by minerals. The organic turns inorganic. Bones become rock.

Only an infinitesimally tiny proportion of life ever becomes a fossil. The odds are vastly against it. 'Assuming an average species-span is approximately equal to between 2 and 5 million years, then the half-billion years of the Phanerozoic Eon has seen the passage of about 1 billion metazoan species. Of those, only something like 300,000 have been described and named – less than one in a thousand,' Zalasiewicz has written. 'Why? Many were soft-bodied, and so unlikely to fossilise, and others were simply rare. Upland terrestrial zones are likely to have eroded, leaving no trace of the animals or plants that once lived there, while the record of the deep ocean floor is obliterated by subduction.'[20]

For a creature to become a fossil, and for that fossil to be one day seen by a human, a series of statistically unlikely events must occur. The creature must die with its body intact. An unusual event such as an exceptionally violent storm must quickly cover it with sufficient layers of sediment, and that sediment must become rock. The sedimentary rock must not be squeezed and changed by pressures and heat deep in the earth. It must then make its way up to the surface so that, some millions of years after the creature died, it is returned to the light. This final stage must take place in an area likely to be frequented by fossilists, during what is so far the narrow sliver of deep time occupied by humans.

In Thomas Hardy's novel *A Pair of Blue Eyes* (1873) Henry Knight, trapped on the side of a cliff, confronts a fossil: 'The eyes, dead and turned to stone, were even now regarding him [...] It was the single instance within reach of his vision of anything that had ever been alive and had had a body to save, as he himself had now.'[21] Thinking about deep time can mean grappling with the shocking sweep of it, the impersonal vast-ness. A fossil delivers a different sort of shock. The shock of the small and personal. The story of a single creature in some sense like ourselves 'with a body to save'. Fossils are the most direct, most evocative evidence that we have for past life in deep time. Where a former landscape must be inferred from the rocks, a fossil is immediately tangible. It can be examined, experienced.

As a child I was taken to look for fossils a few miles along the Dorset coast at Kimmeridge Bay. I remember how the tow-ering, unstable cliffs menaced the beach, and how I found a fossilised ammonite in a piece of shale, and how I put it down, momentarily distracted, and never found it again. Later, my parents bought me a fossil shell from a local gift shop. I still have the shell, but I remember knowing that buying a fossil from the shop wasn't the same as picking it up, rescuing it with one's own hand from the oblivion of the beach and the waves.

Then, walking along the beach at Lyme, I found my second fossil. The shape of the ammonite jumped out at me from the jumble of pebbles and broken, jagged rocks around it: it was the organisation of the shape, the precise lines, the way it was so obviously the product of evolution not entropy.

Ammonites (the name comes from the Egyptian god Amun, depicted with curling ram's horns) were cephalopods, sea creatures something like the living nautilus. The shell, most often curled in a circle, was divided into chambers separated by walls known as septa. As a genus they were extravagantly experimental, diversifying rapidly in shape and size. Ranging in diameter from 20 mm to 2 m, some shells became straight and thin, while others developed spikes or frilled septa. Because they evolved so rapidly, each ammonite species had a relatively short life span, meaning that they make excellent guide fossils for stratigraphers. Travelling back into deep time, ammonites act as waymarkers, used to distinguish intervals of geological time of less than 200,000 years' duration.

I knelt by a rock pool to wash the sand and grit from my ammonite. The central circle of the shell was there with about a quarter of the outer rim. When it was complete, it must have had a diameter of around 30 cm. Because the mineral that has replaced the shell of this fossil was iron pyrite, the outer rim was of burnished gold. It looked man-made. Anthropocentric as the thought is, it seemed incredible that this object was not conceived of and created by a human, for human appreciation.

＊

By early afternoon the sun was shining warmly and the beach getting busy. A father and son took turns to hammer at an ammonite encased in a dark grey boulder. A middle-aged couple picking through the pebbles on their hands and knees looked up as I passed. 'It gets addictive,' the woman said. A man wearing an EU baseball cap took a picture of my gold fossil to show to his wife.

At the Charmouth Heritage Coast Centre, an educational charity housed in a disused concrete factory about a mile down the coast from Lyme, I spoke to the warden on duty, Dan Brownley. A former product design and graphics teacher from Nottinghamshire by way of Dagenham, Brownley also works as a fossil preparator. As a child he used to find fossil ferns in the slates from old mining tips. In his early twenties the father of a friend had a gallery and fossil workshop where Brownley sometimes sneaked off to smoke because it had an extractor fan. 'I'd look at what he was doing and it just fried my brain. I'd look at a rock and then by the end of the week it would be a crab sat there.' He ended up asking if he could learn to prep.

On the desk in front of us were a collection of specimens he was working on. Typically, Brownley leaves his fossils still anchored to the rock they were found in. A smooth, oval stone had been hollowed out, the curve of the stone echoing the curve of the single tawny nautilus sat cupped in the middle. Elsewhere, three pearly ammonites swam across a pale grey limestone backdrop.

Today's preparators have better tools than Anning would have had access to – primarily pens that run off compressed air and which are used to remove the rock from around the fossil – but the basic skills remain the same: 'a lot of patience and nerves of steel'.

Often at first only a tiny fragment of the fossil is visible. Perhaps a spine from an ammonite's shell, or a sort of white crystalline, calcite smear on the grey rock that suggests a fossil buried inside. The preparator must choose where to split and dig in to the rock in order to free the fossil without shattering it. As with a diamond-cutter, the wrong decision could destroy thousands of pounds of material. On his YouTube channel – The Fossil Academy – Brownley, wearing a V-necked T-shirt and blue braces, gives tutorials and tips. Prepping, he says, can't be hurried. In one video he 'pens' down through the layers of rock for four hours (time-lapse footage) – going slowly to avoid damaging the fossil – without knowing whether or not there's actually a decent specimen inside.

'My analogy for fossil prep is to think of a mechanic who's worked on Ford engines all his life,' he told me. 'Then get a Ford engine, cover it in concrete, roll it around, then smack a bit of the concrete off with a hammer so you can see a tiny bit of that engine. If you or me looked at it, we wouldn't know what we were seeing, but if the mechanic looked at it he'd know where everything else was just off of that tiny bit. When you're prepping something you need to know where everything is going to be.'

My own fossil, he said, was not worth preparing. The iron pyrite, though spectacular to look at, was unstable. It would at some point begin to oxidise or rust, finally turning into a handful of dusty fragments.

*

Throughout her life Anning had been subject to gossip and rumours. In particular, there was a persistent story, never proven, that around the age of twenty-one she had become the mistress of a Colonel Birch, a fellow fossil enthusiast more than twice her age. Then, in the 1840s, Lyme began whispering that Mary Anning had taken to drink and opium. This time the rumour was true. She used both substances freely in an effort to dull the pain of the breast cancer that would kill her at forty-nine.

It is a testament to the esteem she was held in by the geological community that at the time of her death she was receiving a special annuity, raised by scientists and the government, in recognition of her work. Though she was not, and as a woman could not be, a member of the Geological Society, their *Quarterly Journal* carried her obituary, written by De la Beche.[22] William Buckland and others subscribed for a stained-glass window dedicated to her memory at the parish church at Lyme. (It shows nothing related to fossils but instead, more appropriately for a Victorian woman, the six corporal acts of mercy.)

In the late nineteenth and early twentieth centuries, interest in Anning declined. Her archive – letters, notes, prayer book etc. – ended up at the Natural History Museum in London, where much was subsequently either given away or destroyed.

'The archive was seen as being "merely" the mementos of a "curiosity", not the repository of an important figure in the history of science, and so a good deal of investigation is still needed,' the historian Hugh Torrens writes. 'One of the most poignant documents I found had accompanied her Common-Place Book [...] to the Dorset County Museum in 1935. It was an earlier letter of refusal by the British Museum, stating how this volume did "not prove to be of sufficient importance" to them. I would have liked to be able to question the judgement that allowed this archive to be so dispersed and so much mate-rial to be "lost".'[23] Here we see the selective forces of history at work. The historical record is wildly incomplete, just as the fossil record itself provides only partial, partially legible details of the former inhabitants of our planet. As a result, there are very few examples of Anning's personal reflections, very few moments where we catch her own voice. Women of her class were unlikely to have had the time or to have been encouraged to write extensive diaries or memoirs. 'Because not that much is known about her personal life, there is a hole and people can kind of either choose to research more and more or can use as a platform to discuss other things – such as a lesbian love affair – that were or were not happening in Victorian society,' Natalie Manifold said.

For a long time Anning's story, if told at all, likely concen-trated on her childhood discoveries rather than her substantial adult achievements and was most often packaged as an inspiring tale for children. But recent years have seen her legacy re-evalu-ated, her life celebrated as that of a pioneering female scientist. There have been biographies and appearances in novels – most famously in Tracy Chevalier's *Remarkable Creatures* (2009). Another film – *Mary Anning and the Dinosaur Hunters*, written and directed by Sharon Sheehan – promises to be a

more conventional biopic and is, at the time of writing, await-
ing release. In London her portrait hangs in the Geological
Society headquarters, and in Lyme locals and visitors leave
shells and fossils on her grave in the parish churchyard.[24]

I suspect that Anning would not have been fazed by her
recent fame. Such was her exposure in her own lifetime that
one day in 1844 the king of Saxony visited her shop. His envoy
recorded the occasion:

> We had alighted from the carriage [...] when we fell in
> with a shop in which the most remarkable petrifactions
> and fossil remains [...] were exhibited in the window. We
> entered and found the small shop and adjoining chamber
> completely filled with the fossil productions of the coast
> [...] I was anxious, at all events, to write down the address,
> and the woman who kept the shop – for it was a woman
> who had devoted herself to this scientific pursuit – with a
> firm hand, wrote her name, 'Mary Annins', in my pocket-
> book, and added, as she returned the book into my hands,
> 'I am well known throughout the whole of Europe'.[25]

10

THE FIRST FOREST

One evening in May 2010 the Devonian plant specialist Chris Berry received a phone call from a colleague at Binghamton University in New York. If you want to see the forest, William Stein said, you need to get over here. A few weeks later Berry, who lectures at the University of Cardiff, had absconded from exam marking and was on the red-eye to Toronto. From there he would take another flight to Albany and then drive around sixty miles to a re-opened quarry near Schoharie Reservoir, just east of the small town of Gilboa in Schoharie County, New York. There, workmen constructing a new dam had uncovered something astounding: the fossilised remains of the world's oldest known forest.[1] Work on the dam had paused to allow the scientists access to the site, but it was only a temporary stay of execution. In a few weeks' time the forest would be covered over once again.

*

I'm fascinated by people who devote their lives to one specific subset of knowledge, whose thinking becomes deep rather than broad, who come to know the world through the prism of baking, say, or car engines. What are the random quirks of fate, the pragmatic or romantic impulses, the formative experiences that cause people to specialise in one area rather than another? Why does a doctor choose to become an authority on the liver, heart or colon? Why does a geologist turn to the Cambrian, the Permian, the Triassic?

Berry was drawn to the Devonian at university. While studying the much celebrated 'Cambrian explosion' – the moment about 540 million years ago when most major animal body plans suddenly appear in the fossil record – 'I looked at the Devonian and I realised that you've also got this other massive explosion of things going on in terms of plants. And I realised that we didn't have any conception of what the world looked like, as it was changing from this time when there were very simple plants to one where there were great big forests.'

The Devonian is the moment in the history of our planet when the world first began to look like *our* world. It is the time when first the plants and then the animals began to move in large numbers from the water to the land. A great greening of hitherto bare rocks, the first footsteps impressed in the soft mud of some long-vanished swamp or shoreline. Named after the county of Devon in south-west England, the period was formally described in 1839 by British geologists Adam Sedgwick and Rory Murchison. It lasted from 419 to 359 million years ago – ending around 100 million years before the appearance of the first dinosaurs – and its most famous rock sequence is the Old Red Sandstone (a group of rocks confusingly not always red or, indeed, sandstone), which, hundreds of millions of years after they were laid down, would be dug up to form, among other things, Hereford Cathedral, St Helen's Chapel at Siccar Point and Tintern Abbey.

Jennifer Clack, Emeritus Professor of Vertebrate Palaeontology at the University of Cambridge and expert in Devonian tetrapods, told me: 'What sparked my interest in the Devonian was Arthur Mee's illustrated children's encyclopaedia. It had sections on the earliest part of the fossil record, and it was always those earlier sections that really intrigued me. Once you got into dinosaurs, it was rather boring because it's all happened, hasn't it?' The world of the dinosaurs would be somewhat familiar to us. There were trees, plants and flowers. There were land-dwelling mammals and reptiles with body plans broadly similar to those seen today. The Devonian was something else. Clack saw pictures of unfamiliar drooping plants growing by the edge of swampy lagoons, strange fish-like creatures hauling themselves out of pools and rivers. 'I remember flicking through the book starting at the beginning and listening to the slow movement of Shostakovich's Fifth Symphony, and if you do that, the music fits the pictures perfectly. That kind of sums up how I imagine the Devonian to be.'

*

'The colonization of land by plants', writes Linda VanAller Hernick in *The Gilboa Fossils*, 'was a series of events nearly as portentous as the emergence of life itself.'[2]

The early Earth was a place of bare, windswept rocks and vast, empty deserts stretching down to the shores of the seas. Without decayed vegetation and active bacteria, no organic-rich soils would have formed. Without the stabilising effect of plant roots, what little soil there was would have washed quickly away when the rains fell. Life existed most often not on dry land but in the teeming oceans.

According to VanAller Hernick, 'it is believed that a kind of scum composed of cyanobacteria and certain types of filamentous green algae was the very first covering of shore and moist near-shore surfaces.'[3] This was the beginning of the colonisation of the land. By 450 million years ago, plants comparable to modern liverworts and mosses may have formed a low, mat-like ground cover in moist areas – something like the baize on a snooker table. Nitrogen-fixing plants and microscopic detritivores followed, breaking down dead organic matter to produce the first nutrient-rich humus soils. We know that during the Silurian (444–419 million years ago), there were centipedes and a plant named *Cooksonia*.[4] Though at most only a few centimetres tall, this is the earliest known vascular land plant, meaning, crucially, that it produced xylem-like tissue, which served to transport water around its body while providing rigidity and some structural integrity. With this adaptation in place, plants would be able to increase their height and expand their range beyond humid environments and wet seasons.

'Then, whoomph, the plants grow taller and taller,' John Marshall, the Southampton-based Devonian specialist told me. 'So you start the Devonian with a few centimetres, and about 30 million years later you're ten metres high. It's explosive.'

Perhaps only someone habituated to deep time would describe a 30-million-year transition period as 'explosive', but certainly, had the plants not developed ways to cope with this difficult movement from water to land, it is unlikely that this world would have become the home of creatures that walk around on two legs and ask themselves such questions as: 'Where do we come from?' 'Why are we here?' 'How is it that our world began?'

*

Berry is one of perhaps a dozen people in the world who specialise in the study of Devonian flora, and his lab in Cardiff houses what he describes as 'the best collection of Devonian tree fossils in the country'. Tall, broad-shouldered and slightly rumpled, he loomed over the miniature Christmas tree next to his office computer. 'It's a Norfolk island pine,' he said. 'Very closely related to the [200-million-year-old] Monkey Puzzle, so it's palaeo-botanically interesting.'

Among shelves of books and piles of paper are pots of brightly coloured playdough, which he uses to make models of the plants he studies. Unlike, say, dinosaur bones, fossil plants are rarely preserved in three dimensions. Often they are a series of fragile black lines of carbon flattened into a piece of rock, impossible to remove from the matrix. 'You have to be able to mentally unsquash them,' Berry explained. Making three-dimensional playdough models is a way of thinking through this 'unsquashing' process.

'Macropalaeobotany is very slow and very hard work. It takes you years to reconstruct a tree because they're big things and they fall apart,' Marshall told me. Without shells or bones, trees are even less likely than other organisms to enter the fossil record; when they do, it is often as unconnected fragments – a branch here, a root or trunk there, and no idea how to piece them together. 'Many attempted reconstructions of the earliest trees,' Berry has written, 'might be described as 'hopeful monsters', not because of their evolutionary status, but because they were chimeras – optimistic constructs of unrelated plant organs joined together to form bizarre-looking plants that only existed in the minds of the palaeobotanists that sketched them.'[5]

When Berry was twenty-three, he spent a year in Liège in Belgium with the renowned palaeobotanist Muriel Fairon-Demaret, examining hundreds of fragments from a group of mysterious Devonian plants known as the cladoxylopsids, the same plants that would be found at Gilboa, and which are now regarded as the Earth's first true trees. Because plant fossils are so delicate, palaeobotanists can't use the sorts of electronic drills favoured by those who work on bones and shells, so Berry had to pick his fossils apart painstakingly by hand, using three-faced needles bought from a specialist leatherworking shop in the city. It was the start of thirty years of study, hunting for cladoxylopsid fossils both in the rocks and in museum and university collections around the world, trying to crack the puzzle of what they looked like and how they grew.

Pinned on to the wall of Berry's office are some of the visual reconstructions that have accompanied his papers. While he spoke to a student who had come to collect an essay, I stood and looked at them. When you see a reconstruction of *Calamophyton primaevum*, a type of cladoxylopsid, your brain says *tree – but not quite*. Perhaps this is what it would have been like to be a seventeenth-century European arriving in the Pacific and seeing palm trees for the first time, or an eighteenth-century Polynesian viewing your first British oak. *C. primaevum* is at once familiar and strange. A bulbous swelling at the base of the tree thins into a long, slender trunk, most of which is covered with what look like tiny spines that mark the places where branches had grown and then fallen off. The extant branches are all clumped together at the top, giving something of a palm-tree effect or, as I read it described elsewhere, a stick of celery. Instead of leaves, the end of each single branch splits into five or six smaller branches, like fingers on the end of an arm.[6]

Opening an old-fashioned wooden collector's cabinet, Berry took out a flat, disc-like cladoxylopsid fossil, around the size of a large dinner plate. 'When I first looked at this,' he said, 'it blew my mind.'

The fossil was a glassy pale grey circle, around 2 cm thick and 70 cm in diameter. Around the edge of the circle was a thick band of dark oval shapes, large and small, something like the markings on the coat of a leopard. It was a nice fossil, I thought, staring at it and experiencing one of those uncomfortable moments when you know that your appreciation for something is not going to match the enthusiasm of the person showing it to you. I muttered something about it looking like a great fossil.

Berry had to explain. Here was a cross-section of a cladoxylopsid trunk perfectly preserved in silica from volcanic eruptions. The glassy silica had filled the plant cells but not destroyed the cell walls: every detail was as clear as if Berry had time-travelled back to the Devonian, cut down the tree himself and sliced it into sections. With this fossil he was able for the first time to see the outrageously strange interior of one of the Earth's earliest trees.[7]

When I asked Berry what made someone in his field successful, he said that it was a result of acquired knowledge and luck. This fossil was an example of luck. It comes from the deserts of Xinjiang autonomous region in north-west China. To date, no other locality has produced such specimens. Were it not for his long friendship with a Chinese palaeontologist called Hong-He Xu, from the Nanjing Institute of Geology and Palaeontology, it is unlikely that Berry would have had the opportunity to study it. In the past he had visited Xinjiang himself, but today the area is effectively closed to Western scientists. It is the site of what the Chinese authorities describe as (re-)education schools for Xinjiang's Muslim minority, and what the BBC and other news sources call internment camps.

Berry and Xu named the tree *Xinicaulis lignescens*.[8] *Xin* means 'new' in Mandarin (and alludes to Xinjiang province), *caulis* is the Latin for 'stem' and 'lignescence' means 'becoming woody'. Naming things, Berry mused, sounds glamorous but is actually incredibly tedious. 'It involves trying to think of something that is significant, sounds nice and hasn't been used before.' Naming a plant or dinosaur after yourself is considered bad form. Romantic partners are also out – bad luck, apparently – though naming something after a colleague is OK. Berry and Xu chose *Xinicaulis lignescens* because, to the researchers' amazement, this fossil demonstrated 'an entirely new way of becoming a plant'.

A tree can only grow so high before it needs to expand its trunk in order to stop itself from breaking or flopping over. Modern trees deal with this problem by growing outwards as well as upwards, and to do this they produce annual rings of wood that slowly expand the circumference of the trunk underneath the bark. Some early trees such as *Archaeopteris* used this strategy. *Xinicaulis lignescens* did something completely different and much more complicated.

Under the microscope, you could see that each dark oval inside the trunk was composed of many tiny concentric circles. Unlike a modern tree, *Xinicaulis lignescens* grew its woody material in individual strands, each one acting like a self-contained miniature tree and all joined together by an intricate network of interconnecting woody tissues. The space between the individual 'trees' was filled with soft tissue, and most of the inside of the trunk was completely hollow. The resulting structure must have been something like a wooden Eiffel Tower.

To allow the tree to expand outwards, the interconnecting woody tissues were slowly ripped apart as the tree grew. But to allow the tree to remain structurally sound, the damage needed simultaneously to be repaired, meaning that a *Xinicaulis lignescens* was in a state of continual, controlled internal collapse. 'It's just insanely crazy,' Berry said, shaking his head. No tree does anything as complicated today. 'Maybe the lack of competition from other tree types at the beginning was what allowed this to occur,' he has speculated.[9]

*

The fossil trees at Gilboa have been known (as individual specimens rather than a forest) since at least 1871, when a description of the sandstone cast of a fossil tree stump was published in the *Quarterly Journal of the Geological Society*.[10] It was the first recorded discovery of a fossil tree in North America.

Founded in 1848, the town is named – inauspiciously, one would think – after Mount Gilboa, a biblical site in Israel where King Saul's sons were killed by the Philistines and Saul took his own life. In 1917 the local creek was dammed to form a reservoir for the city of New York, and the settlement of Gilboa was moved to its current position. As quarrying for the dam progressed, more and more fossil stumps were uncovered – some with an immense circumference of over 3 m, the stump flaring out at the base like an amputated elephant's foot. From 1921 Winifred Goldring, state palaeontologist of New York, worked at the site, producing the first and largest coherent body of work concerning these early trees. Then, in 1926, the quarry was backfilled with waste from the dam construction. The rescued fossils found their way into various museums and the occasional local back garden.[11]

The story of the Gilboa forest might easily have ended there. But in 2009 the east coast of America was hit by storms and flooding, and when the dam showed signs of movement, local officials decide it was time to rebuild. Material was needed to make road bases for the lorries used in construction, and so the old quarry was reopened and the backfill removed.

Stein and Frank Mannolini, from the New York State Museum, were tasked with supervising work at the quarry. At first they concentrated their efforts on a section around 3.5 m above the quarry floor, where tree stumps had previously been found, but then they noticed some strange mounds on the quarry floor itself, each with a depression in the centre.

When Goldring was there, Berry explained, 'it was a working quarry with crap everywhere, so she'd never have seen the bottom – or only little bits of it.' This was uncharted territory. Each mound, Stein and Mannolini realised with growing excitement, was the site of an ancient Devonian tree. In effect, what they had before them was the first ever map of a Devonian forest.

*

By the time Berry stumbled, jet-lagged, into the quarry site, an area of around 1,200 m^2 had been hosed down to expose the bare rock. Before now, illustrations and museum dioramas of Devonian landscapes tended to show the trees and plants as isolated specimens within what Berry calls 'Devonian botanical gardens'. Now the scientists had the first direct evidence of early forest ecology.[12]

'You spend twenty years trying to work out what these trees are, and then you get walked into this quarry first thing in the morning after this horrendous journey, and I can remember that I sat there with crossed legs looking around and it was remarkable,' Berry said. 'It wasn't just in my head any more. For the first time I had the experience of actually sensing this ancient environment.'

As he sat there, it was as though the ancient Devonian forest stirred to life around him. A warm, tropical place, perhaps at the edge of a shallow estuary, and, for the solitary visitor, a strangely silent one. In the Devonian there is no birdsong, no animals crashing through the undergrowth, only the soughing of the wind in the canopy, the sigh and clatter of branches and the splash of a fish coming up to the surface, leaving behind an expanding circle of ripples in the oily water. Tightly packed, slender-trunked cladoxylopsid trees stand 8, 10, 12 m tall, erupting at the very top into a plume of branches frilled with fine, leaf-like filaments – as many as 900 to 2,100 trees in an area the size of Trafalgar Square.[13] Between them, the woody stems or rhizomes of aneurophytalean progymnosperms crawl along the forest floor. Up to 15 cm thick and as long as 4 m they twist and coil snake-like around the trunks of the upright trees.[14]

Based on their research, Berry and his colleagues have produced a map of Devonian Gilboa, allowing armchair travellers to trace their way among the trees of one of our planet's first forests, just as those forests were beginning to change the Devonian world in dramatic new ways. Then in 2020 Berry published a paper on a second ancient forest that has been discovered in a sandstone quarry near the town of Cairo, some 40 km east of Gilboa. Dating from around 386 million years ago, it is around 2 or 3 million years older than Gilboa and is, at the time of writing, the world's oldest known forest.[15]

'Changing the Earth's soil is the biggest thing the new forests do,' John Marshall told me. As tree roots broke down rocks to release minerals and nutrients, more rich, stable soils were formed, which in turn encouraged further plant growth, slowing rates of erosion and reconfiguring the shape of the land. Thriving trees profoundly altered the water cycle (for instance, reducing run-off and increasing precipitation). They removed CO_2 from the atmosphere, which some researchers claim led to the 90 per cent decline in CO_2 levels through the Palaeozoic, ultimately triggering a period of global cooling and what Berry has called 'the first ice age on a forested Earth'.[16] Where the newly nutrient-rich soils were washed into the ocean, the excessive growth of algae resulted in the depleted oxygen levels in the water. Other researchers have speculated that this may even have caused the cluster of extinction events at the end of the Devonian.[17] And of course it is the remains of ancient trees – those growing some millions of years later in the Carboniferous – that coal miners would dig up to fuel the Industrial Revolution across Great Britain, Europe and the United States, beginning the process that would lead to our twenty-first century climate emergency.

But all that is to come. At the beginning, this new world is full of life: striving, growing, greening. Centipedes, spiders and giant millipedes make their way through fallen cladoxylopsid branches piled up on the forest floor.

The Earth is half as close to the moon as it is today, and at night this larger, brighter moon looms enormous and golden. It gleams on the branches of the trees, on lazy bubbles bursting in the swampy water, on the dark, flat surface of the shallow estuary where, on the sandy floor, a fish-like creature with the flattened head and nostrils of a crocodile uses its fins to propel itself forward, through the sand and mud and reeds.

This creature has never walked on dry land before, but soon, following the journey of the plants, it will make its way towards the water's edge. Over time, fins will become limbs. The first land animals will appear.

11

WHAT WE TALK ABOUT WHEN
WE TALK ABOUT DINOSAURS

London ... Implacable November weather. As much mud
in the streets as if the waters had but newly retired from
the face of the earth, and it would not be wonderful to
meet a Megalosaurus, forty feet long or so, waddling like
an elephantine lizard up Holborn Hill.

Charles Dickens, Bleak House

One spring afternoon a group of mammals went bowling,
dressed up in reptile costumes inspired by the film *Jurassic
Park*. We were celebrating an upcoming wedding. The groom
had been given a velvety green head and velvety mittens in the
form of large claws. There were sexy dinosaurs in leotards,
dinosaurs with neck frills made of cardboard, dinosaurs with
long, unruly tails that kept knocking the drinks off the tables. I
had gone as the park, wearing a jumpsuit decorated with jungle
foliage and a necklace made from plastic toy dinosaurs. It was
the second wedding celebration with a dinosaur theme that I'd
attended. The third if you count Godzilla.

Jurassic Park was released in 1993. Such was its success it can probably claim to have single-handedly produced a huge new boost to the popularity of palaeontology. 'The "Jurassic" generation is 100 per cent a thing,' palaeontologist Joseph Frederickson, a professor at Southwestern Oklahoma State University, said in a recent NPR interview. 'I have colleagues – so many colleagues – who [...] were just children when *Jurassic Park* came out. And I have no doubt that it meant just as much to every one of them as it did to me and can really point to that moment as being one of these life-changing events that really made them want to get into palaeontology.'[1] It's a film that encapsulates the melancholy longing at the heart of palaeontology. That dry bones be re-clothed in muscle and flesh. That the long-dead should walk again.

When it came to talking about dinosaurs, the scientists, artists and curators I met were mostly divided into two camps. There were the fanboys, those who had been passionate about dinosaurs ever since some fondly remembered childhood encounter. The others seemed faintly embarrassed by it all. 'It's the science that interests me,' one palaeontologist stressed; 'the dinosaurs are just a medium for me to do science.' Jakob Vinther, a palaeobiologist at the University of Bristol, worried that 'Sometimes I feel a little bit bad working with dinosaurs because dinosaurs are opium for the people in some ways. Fox News loves them, and you can detract from really serious problems in the world by saying stupid things about dinosaurs. I mean, yes, its cool, but the amount of interest scientists studying dinosaurs get is ridiculous.' His main research involves prehistoric invertebrates, and recently his team discovered the ancestral mollusc. 'We can show what the ancestor of all molluscs looked like but it hardly made the news!' he said, looking mystified. 'Journalists are just like, no, we want dinosaurs. Dinosaurs are fun.'

*

Because dinosaurs are fun, I went to the Natural History Museum in London with one of my brothers, his wife and my five- and seven-year-old nephews. It was a rainy Saturday, and the museum was full of children and their attendant adults. Many were heading towards the dimly lit dinosaur gallery on the ground floor, where bony, eyeless figures loomed out of the gloom.

Today non-avian dinosaurs – extinct for approximately 66 million years – are everywhere, so familiar that it's easy to look at a dinosaur without really seeing it, and difficult to comprehend how awesomely strange these colossal monsters must have appeared to the men and women who first dragged them out of the Earth. I stood in front of one of the specimens. Sixty-eight million years ago this creature died in the mud, its bones changed slowly into stone. Now someone has pulled them out of the earth; now they have been reassembled in the dim light of the museum. There were fist-sized eye-sockets in the eyeless skull. Two jagged rows of teeth. A curving grey rib cage. Two grasping hands.

But the slathering jaw was strangely at odds with the hands, which looked so familiar, almost homely – almost, I couldn't help thinking, human. We share with the dinosaurs a basic body plan; our ancient common ancestry can still be seen in our skeletons. Like them – perhaps the basis of our fascination – we are apex predators. Like them, we are – probably – ultimately at the mercy of environmental forces beyond our control.

As a child I loved going to see the Victorian dinosaur statues in Crystal Palace Park in south London – deliciously, alarmingly large and strange but safely out of reach both physically on their islands in the park's man-made lakes, and temporally.

Often a childhood encounter with a dinosaur marks the first attempt to engage with the immensity of deep time. The last non-avian dinosaur died around 66 million years ago. As a genus, non-avian dinosaurs existed for more than 150 million years.[2] Our own genus – *Homo* – has so far lasted for only about 2.5 million, modern humans a scant 200,000. Though the lifespan of an individual dinosaur was almost certainly short, compared with *Homo sapiens*, dinosaurs as a genus are impossibly long-lived. To put it another way, *Tyrannosaurus rex* (which lived 68 to 66 million years ago) didn't make an entrance until around 77 million years after *Stegosaurus* died out, meaning that *Tyrannosaurus* is closer in age to us than to *Stegosaurus*.

Both my nephews like dinosaurs, but the younger one is the real enthusiast. He currently owns more than 100 plastic dinosaurs, from small, anatomically detailed Papo figurines to a large rubbery *Tyrannosaurus rex* from Toys 'R' Us. Since he was about two he has been aware in some way that the plastic dinosaurs represent real creatures that lived and died and no longer exist, having been wiped from the face of the earth for ever. This knowledge doesn't seem to worry him unduly. He is, however, a stickler for realism: any attempts to introduce anthropomorphic elements into his dinosaur games are sternly rebuked.

'Dinosaurs can't talk,' he explains again patiently.

'What can they do?' I ask.

'These ones are plant eaters, they go over here; these ones are meat eaters, they go over here.'

'And then?'

'And then these ones kill these ones.'

Familiar with the museum, the boys led the way confidently into the maze of exhibits. Collections of bones reared up out of the darkness. Fearsome horns. Brown teeth. Clasping claws. *Scolosaurus!* they called out. *Parasaurolophus!* Like other primary school – and pre-school – enthusiasts I have met, they can list a growing number of complex Latinate names. The younger one, all plump pink cheeks and a sweet smile, favours the more violent dioramas: 'You know a *Triceratops* could actually kill a *T. rex* by stabbing him in the thigh bone,' he told me approvingly.

Outside the dinosaur gallery is an entire gift shop devoted to dinosaur-themed stuff for kids. In the twenty-first century, anything a child could conceivably want or need now comes either shaped like, or patterned with, dinosaurs. The first page of Google Shopping offers, among other things, a pterodactyl lampshade, a dinosaur bracelet, a grow-your-own-dino kit. And the books. Dinosaur-themed books include: *The Dinosaur that Pooped a Planet*, *Mad About Dinosaurs!*, *Help, My Dinosaurs Are Lost in the City* and *The Dinosaur Potty Training Book*.

'That fact that they can be big and scary is really important,' child psychologist Laverne Antrobus told me. 'Learning about the limits of life, being scared, all of these things can be explored through dinosaur play.' Thinking about dinosaurs has a *frisson* of the real not experienced with, say, dragons, but they are still safely dead and in the past. There is no way a dinosaur can hurt you. 'It's terror you can play with; dinosaurs stimulate imagination and playful aspects of risk that children use to learn about the world.' My nephews, for example, recognise dinosaurs as scary but, with the exception of the really terrifying bits in the *Jurassic Park* films, are not actually scared by them. Dinosaurs don't seem to induce the same night terrors as, say, the bogeyman under the bed.

But what's also important, Antrobus thinks, has less to do

with dinosaurs themselves and more to do with knowledge acquisition. 'Often this knowledge feels like the first really big stepping-stone into the adult world. You've memorised all these dinosaur names and reeled them off, and you get a whoop from adults, particularly your parents, who I suppose start to think about your academic capacity, your processing skills, because after all this is incredible: before you can even read phonetically you can say bronchi-whatever. It gives real confidence. I think it's also probably one of the first packages of knowledge children get.' They are getting high, basically, on discovering that there's this certain subset of information that they can learn and master and add to. 'You start to synthesise and you think about what you know in different ways. It's a chain of events: you start with the names, then what they look like, what they ate, whether or not they got along.' And for the adults watching the child it is incredible; we praise the child and reinforce that this is something good and special that they can do.

Some children will go on to become palaeontologists or life-long amateur rock hounds, but for many the period of intense interest begins to wane by the age of eight, if not before. Perhaps at school they meet other dinosaur fanatics and realise that having this knowledge is not actually so special, and begin to lose interest. Or they intuit that with schooling comes a require-ment to develop a different, broader skill set. Praise from adults no longer comes predominantly from your ability to recite the animals of the Lower Cretaceous – it's about making friends, mastering maths and spelling and reading, taking part in sports day and the choir and the nativity play.

In the museum, the five-year-old spotted something. *Deinonychus!* Look behind you! Helen! *Deinonychus!*

Behind me were two life-size, animatronic feathered dino-saurs, one with a bloodied mouth. We watched them for a while, and I asked him why he likes dinosaurs so much. He thought for a moment. 'I like them because they are big and scary and they come in different colours and they're all different.'

A little further on I came across a glass and wood cabinet containing two toffee-coloured teeth and something like the horn of a rhinoceros but actually a detached thumb spike. These fossils were discovered in 1822 by the doctor and naturalist Gideon Mantell.[3] He decided that they came from an entirely new creature, which he called an Iguanodon, meaning 'Iguana-tooth', because the teeth were like those of an iguana. In 1833 Mantell published a description of a creature he named *Hylaeosaurus*.[4] Eight years later the British anatomist Richard Owen studied these creatures and one named *Megalosaurus*, described in 1824 by William Buckland, the famously eccentric professor of natural history at Oxford.[5] Owen noted that they shared features of the hip lacking in other reptiles, and that they were very, very large. He named them *dinosaurs*, a name meaning something like 'terrible reptiles', but with 'terrible' intended to mean 'awesome' or 'fearfully great'.[6]

The newly named dinosaurs were an immediate public hit. Owen went on to become the superintendent of the natural history collection in the British Museum. His reputation as a brilliant anatomist has been tainted by a concurrent reputation for meanness and scientific skulduggery; at one point he was accused of claiming for himself the work of other scientists (though he seems to have been supportive of, for example, the work of Mary Anning). In particular, he feuded with Mantell and, according to some sources, prevented the publication of some of Mantell's papers. Mantell himself neglected his medical practice for palaeontological studies and consequently sank into debt. Severely injured in a carriage accident on Clapham Common, he began self-medicating with opium and died of an overdose in 1852. Owen had a portion of his rival's spine pickled and displayed in a vitrine in the Royal College of Surgeons.

In 1881 Owen was instrumental in the decision to open a new National History Museum, and became the institution's first director. Not only had he invented the dinosaur, but he had also found a way to allow the general public to share in his work, to make the science of palaeontology accessible to larger numbers than ever before. In 1969 Mantell's spine was taken from the shelves of the Royal College of Surgeons and destroyed. Reasons of space were cited.

*

In June, Jonny and I flew to Salt Lake City and then drove south past strip malls and fast-food franchises. At a Subway by a busy intersection in the city of Provo we pulled over and tried to buy a coffee. The women behind the counter glanced at each other – they'd never used the coffee machine. In Provo, which is 88 per cent Mormon, there isn't much call for it. Later, checking the map, I traced our destination on the freeway into the heart of the state of Utah. Somewhere to the east was the reason I had come there: one of the country's newest National Monuments – signed into being in March 2019 – which was also the densest deposit of Jurassic dinosaur bones to be found anywhere in the world.

The American west is dinosaur country. By the late 1870s, North America had become the place to hunt for fossils. In what is sometimes referred to as a 'dinosaur rush' and sometimes as the 'bone wars', dinosaur hunters headed west to Montana, Wyoming and Utah. It was during this period that some of the best-known dinosaurs were discovered, *Tyrannosaurus*,

Triceratops, Diplodocus and *Stegosaurus* among them. In particular, two men – Edward Drinker Cope and Othniel Charles Marsh – competed to dig up bones and ship them back east to the great museums and universities, naming over 130 dinosaurs between them.[7] Living up to the 'Wild West' moniker, and spurred on by the men's bitter rivalry, their dinosaur-hunting crews spied on one another, dynamited bone beds to stop their opponents from digging there and occasionally stole each other's specimens. Later, the two men devoted much energy to savaging each other in the press, alleging bad science, theft and underpayment of staff, with the result that for some years the entire funding for palaeontology in the USGS was withdrawn.

Like Mantell, part of Cope eventually ended up in a museum. On his death-bed he issued a challenge: his brain, he wagered, was larger than Marsh's. To prove it, he would donate his body to science and have his skull measured; when Marsh died, he was to do the same. Marsh never took up the challenge; Cope's skull was stored in the University of Pennsylvania Museum of Archaeology and Anthropology, now usually known as the Penn Museum. Unlike Mantell, he still resides there today, preserved in a velvet-lined box, catalogued and organised just like one of his beloved dinosaurs.

Public interest in dinosaurs continued for some time, but during the early twentieth century, as Darren Naish and Paul M. Barrett point out in *Dinosaurs*, they fell out of favour.[8] The general opinion was that 'mammals (especially those belonging to modern groups, like rodents and horses) were more deserving of study than dinosaurs', which were characterised as slow-moving, unintelligent, cold-blooded, 'uninteresting with respect to our understanding of life on Earth as a whole, and generally not all that worthy of attention'. In America, First World War protesters constructed a papier-mâché, heavily plated *Stegosaurus* – a jibe at those who preferred warfare to diplomacy. As protestor Walter G. Fuller wrote, 'These beasts, all armor-plate and no brains, had no more intelligent way of living than that of "adequate preparedness" […] piling on more and more armor, until at last they sank by their own clumsy weight into the marsh lands.' [9] Dinosaurs were viewed as animals that had grown too large, too lavish, for their own good. In business jargon 'dinosaur' became (and still is) a term for large companies that have failed to change with the times, and will eventually crumble in the face of smaller, more nimble businesses. Elsewhere we continue to use the term 'dinosaur' to refer to something outdated, as in the film *GoldenEye* (1995), where Judi Dench's M tells Pierce Brosnan's Bond that he's a 'sexist misogynistic dinosaur, a relic of the Cold War'.

Dinosaurs would spend the next thirty or so years relegated to stories for children and fodder for screwball comedies such as Howard Hawks's delightful *Bringing Up Baby* (1937), about a hapless palaeontologist (Cary Grant), a Brontosaurus, a leopard and a free-spirited society girl (Katharine Hepburn). It wasn't until the '60s that they would find favour with serious scientific communities and the general public once more. And when they did, the Cleveland-Lloyd Dinosaur Quarry in Utah would quickly become a magnet for a new generation of researchers.

*

No one knows who first discovered the bone bed, but palaeontologists have been digging there since the 1920s. 'It truly is a world-class site,' says Joseph Peterson of the University of Wisconsin Oshkosh. 'Beyond anything I've ever worked on before.'

To reach the bones from the nearby town of Price, you go

south 12 miles and then turn on to an unsurfaced dirt track that winds for 13 miles through a semi-arid landscape of yellow rocks and scrubland – and as you travel further away from Price, you also travel back in time, the landscape becoming progressively older, from the Cretaceous, 80 million years ago, in Price until you reach the 146- to 156-million-year-old Jurassic Morrison Formation. During the Jurassic, the Cleveland-Lloyd site would have been situated in the middle of a vast, flat plain with scattered lakes and sluggish, muddy rivers that meandered from horizon to horizon – something closer to the landscape of modern-day east Africa.[10] Today the palaeontologists are excavating an area next to a low, rubbly hillside. Distant orange buttes shimmer in the heat, and early in the morning when we arrived, the only sound was the flapping American flag up on a pole blowing in the hot desert wind. Two corrugated-roofed sheds had been set up to protect the ongoing dig. Blackened bones emerging from grey soil; things long hidden coming out into the light.

For the last twenty years Michael Leschin, of the Bureau of Land Management, has been in charge of the Cleveland-Lloyd quarry. Grey curly hair twisted into a ponytail underneath a cream, sweat-stained cowboy hat, he mostly works alone manning the visitors' centre and trying to work out ways to develop and conserve the site without increasing his budget. When we met, the quarry had not yet been granted National Monument status, about which Leschin was dubious. Would it bring an influx of visitors without a concurrent influx of additional funding, straining the limits of the site's few resources?

'In 1960,' Leschin said, 'William Stokes, working for the University of Utah, started what became known as the Cooperative Dinosaur Dig.' Stokes knew that there were a lot of bones at Cleveland-Lloyd, but he didn't have the money to get them out. It was fortunate, then, that the 1960s coincided with the beginning of the so-called 'dinosaur Renaissance'.

Led by John Ostrom of the Peabody Museum of Natural History at Yale University, new scientific advances saw a whole-sale reimagining of the dinosaurs. No longer slow, cold-blooded and dull evolutionary dead ends, the dinosaurs, Ostrom argued, were actually a phenomenal success story.[11] He recast them as smart, hot-blooded, evolutionarily superior creatures, and his ideas were later vividly (some say too vividly) brought to public attention via the writings of his former student Robert T. Bakker – in particular the non-fiction book *The Dinosaur Heresies* and the novel *Raptor Red*, written from the point of view of a female *Utahraptor* and chronicling her struggles to survive in Cretaceous-period America. (The character of Dr Robert Burke, in the Jurassic Park sequel *The Lost World*, was apparently inspired by Bakker.) The '60s were swinging, and dinosaurs were suddenly hip again.

Stokes began phoning museums offering dinosaur skeletons in exchange for either money or workers. (Thanks to a law written primarily to protect against the removal of archaeo-logical resources, Stokes's actions were actually illegal, but no one seems to have complained.) To date, more than 12,000 bones have been recovered from the quarry, and dinosaurs from this small patch of Utah can be found in sixty-five different institutions around the world, including Edinburgh, Liverpool, Kuwait, Milan, Turkey and Tokyo, where the quarry provided the first dinosaur ever to be exhibited in Japan. So many bones still remain that palaeontologists working on the site today are constantly in danger of stepping on new finds. 'It's the only site in the world where you can get frustrated by finding a new bone, because they're just everywhere,' one says in a promo-tional video for the site.

Because Cleveland-Lloyd has been excavated on and off since the 1920s, the history of the site is also the history of doing dinosaur science. When palaeontologists first started working at the quarry, piecing together dinosaurs was the main business. 'They didn't even bother mapping the bones here back then,' Leschin said. Today palaeontologists want to explain the dinosaurs in the context of their environment. They might be said to be interested in narrative – the story of how the dinosaurs ended up where they did, the story of evolution that they are part of, the story of the ecosystem they lived and died in. And when it comes to narrative, Cleveland-Lloyd is a 148-million-year-old detective story. No one knows how all these dead dinosaurs ended up in one place – and why there is an unheard of three-to-one predator-to-prey relationship. 'It's like, come on! That doesn't happen!' Leschin said. Typically you'd expect prey to greatly outnumber the predators, but at Cleveland-Lloyd more than 75 per cent of the bones recovered belonged to one of the most terrifying of the carnivorous theropods (often known as the 'predatory' or 'meat-eating' dinosaur group): *Allosaurus fragilis*. So many of these have been discovered at Cleveland-Lloyd that palaeontologists have been able to construct almost an entire Allosaurus life-cycle using the bones. A lithe predator that looks, to the uninitiated, a little along the lines of *Tyrannosaurus rex*, *Allosaurus* could grow to a length of 12 m (roughly as long as a double-decker bus), weighed as much as 2,000 kg (about the same as a Land Rover) and was equipped with three wickedly curved claws on each hand and a mouth generously filled with serrated teeth.[12]

Numerous hypotheses have been proposed to explain the origin of the bones in the quarry. Were the dinosaurs poisoned? Did they die because of drought? Were they trapped in thick mud? Was this, perhaps, a seasonal ephemeral pond, into which dead bodies were washed before being covered up by sediment? Did the water then turn toxic, explaining why there are almost no bite marks on the bones? 'We have a lot more questions than we have answers,' Leschin said. 'Back when dinosaurs were new, just finding out they existed was amazing. But now it's like, huh, how did they all get here?'

The site is currently being excavated by the University of Wisconsin and the University of Indiana. Leschin showed us the main dig site – a shallow pit inside one of the corrugated sheds that, out of necessity, doubles up as Leschin's storage unit, with boxes of tools, lengths of wood and coiled ropes piled up around the sides. Here Leschin spends a lot of time thinking about how, without spending much money, you can stop water getting underneath the walls of the shed and damaging the bones. We looked down into the pit. Several dark bones like black gnarls of tree roots were visible sticking out of the grey earth. Some had been covered with coats of white plaster of Paris to protect them when they are extracted.

During the dig season, the pit will be filled with students and professors squatting in the earth, chipping and brushing away at layers of rock – a painstakingly slow business. For the first time, photogrammetry is being used to map the site. 'We're going through the quarry in 10- to 15-centimetre levels. Every time we get down to that level we map our bones on a grid system. We then take a number of photographs to produce a photogrammetric model, and then we'll have a 3D model on the computer of that level of excavation and can study how that deposit changed over the years,' Peterson explained in an article for the *Smithsonian* magazine.[13] By studying these models, he hopes to see whether the dinosaurs at the site were felled by a single catastrophe or whether the bodies accumulated over time.

Other studies include analysing the quarry's geochemistry and running experiments to find out what happens to theropod carcasses when they're soaked in water, approximating the conditions of Jurassic Cleveland-Lloyd. In the absence of any fresh *Allosaurus* carcases, the scientists are using birds.

'My approach to the quarry is the same as a crime scene or archaeological site – leave no stone unturned,' Peterson says.[14]

*

On a Friday afternoon the hot, wide high street of Price, capital of Carbon County, Utah, was almost empty, many of the shops closed – some taking off for an early weekend, perhaps, some that looked more permanently shuttered. Inside the coolness of the Utah State University Eastern Prehistoric Museum, across the road from the district court, some of the creatures from Cleveland-Lloyd have come to rest.

Kenneth Carpenter, the museum director, showed me the dinosaur hall, where a *Camptosaurus*, with a skull like a horse, faced off against a rearing *Allosaurus*. Carpenter, a short, dark-haired man wearing turned-up blue jeans, speaks in a quiet voice and seems permanently privately amused about the world around him. 'We've tried to incorporate as much real bone as we can,' he said, pointing to the display. 'The Smithsonian, for reasons only they understand, have taken down most of their real stuff and put up casts. To me, that would be like going to the Louvre to see a replica of the *Mona Lisa* and not the original. Even though it might look the same, it is psychologically not the same. I don't care what anybody tells me: people want to go to museums to see the authentic real stuff. They don't want to see a replica.'

Where Jakob Vinther seemed faintly embarrassed by his

dinosaurs, Carpenter was enthusiastic. Born in American-occupied Tokyo in 1949, when he was five years old his mother took him to see *Godzilla*. It did for him what *Jurassic Park* did for the younger scientists: 'I was so wowed that I decided right then that I was going to be a palaeontologist.'

The bones at Cleveland-Lloyd have helped him solve some of the questions that keep dinosaur experts awake at night. Questions such as 'Would an *Allosaurus* have bested a *Stegosaurus* in a fight?' And 'With what force would a *Stegosaurus* need to attack an *Allosaurus* in order to cause a puncture wound?'

'I've been asked before, if I had a time machine and could go back and see what it was like, would I want to? My answer has usually been no, and the reason is that for me the fun of palaeontology is trying to figure it out. It's mostly a mind exercise. If I had a time machine and could go back then the mystery has gone. It would no longer be palaeontology but biology. And yeah, it's nice to go to the zoo and see the animals but I don't necessarily want to study them.'

At the museum, Carpenter is attempting to tell the grand sweep of the deep time story of eastern Utah, which is, of course, much larger than the span of the dinosaurs' reign. But it's the dinosaurs, you sense, that have his heart or whatever part of us it is that stays constant from childhood. In a recent book – *The Carnivorous Dinosaurs* – Carpenter examines the puncture wound found on a metatarsal bone from a Cleveland-Lloyd *Allosaurus*, speculating that it was caused by the spike on a *Stegosaurus* tail. Sounding like a forensic pathologist in a crime show, he writes that: 'On the basis of an empirically determined best fit, the spike pierced the transverse process from 58° below the horizontal plane, 33° anteriorly from the vertical transverse plane, and 10° laterally from the sagittal vertical plane ... The spike apparently did not exit cleanly from the wound … thus widening the hole.' He goes on to speculate that the fan of upright plates on a *Stegosaurus*'s back 'may have served to defeat attacks aimed at tracheal crushing' – perhaps the *Allosaurus*'s preferred attack move.[15] Considered one way, I can't help thinking that this sort of science is at heart not too far removed from my nephew's 'You know a *Triceratops* could actually kill a *T. rex* by stabbing him in the thigh bone'.

I asked Carpenter a question that I've often wondered about palaeontologists: Why bother spending years of your life reconstructing the appearance and behaviour of long-dead creatures, a pursuit with no practical relevance to our world today?

'I want to educate people, hopefully give them a greater appreciation of past life, and that the present is very different from the past,' he said. Another palaeontologist, who asked not to be named, was perhaps more candid. 'My personal opinion? I don't care. I'm just having fun doing it.'

'I think the main reason for natural science broadly is to understand the world surrounding us and understand where it's going,' the Swedish palaeontologist Johan Lindgren said when I put the same question to him. 'And as we cannot see the future, the only data we can gather is from the past. If we want to understand what is going on today, we need to look back in time.'

'If you want to be cantankerous, well, it's not going to cure cancer or solve any energy crisis,' Vinther said. 'But I think generally that we should have some natural obligation to understand why the world that we live in is the way that it is – and understanding how things got assembled through deep time is an important part of that. Because if we don't understand the things around us then we don't understand why we should take care of them. And then we might as well dump plastic in the oceans and not care if all the red squirrels disappear. Also, I think there's a natural curiosity for trying to understand these things on a par with *X Factor* or going to art museums. I mean, why should we care about art? Why should we care about *X Factor*? If people are interested in it, then it's worth doing.'

*

In 1997 the Field Museum in Chicago paid $8.3 million for Sue, the fossil of a *Tyrannosaurus rex*.[16] (Funds had been raised by the California State University system, Walt Disney and McDonald's.) At the time of writing it is the most expensive dinosaur sale to date.

'*Jurassic Park* gave the market a big shot,' James Hyslop, Christie's science and natural history expert, told me when I visited the auction house in St James's, London. 'I don't think there would have been the millions for Sue if *Jurassic Park* hadn't happened.'

By the millennium, dinosaurs had been reinvented once again – this time as status symbols for wealthy individuals and companies. Leonardo DiCaprio, Russell Crowe and Nicolas Cage are all known collectors. The skull of a *Tyrannosaurus rex* sits in the lobby of a Californian software firm.[17] Other bones end up as interior design pieces in the rooms of grand houses and apartments. Carnivores are more popular than herbivores. 'Lazy as it is, there's probably some truth in the stereotype of this being an alpha male purchase. People want something big and ferocious that will chomp up their neighbours,' Hyslop said. 'And if you want something really exciting and snazzy, you're always going to want a *T. rex*.' With a *Tyrannosaurus rex* tooth Hyslop expects to make £5,000 to £10,000 every time. With an Allosaurus tooth – which is older and rarer – he'll struggle to make £1,000. 'It all comes down to brand names. Part of that is because it's a fairly young market – people aren't hugely educated about it – but also it's because that's how the art market operates.' People are more willing to buy a *Tyrannosaurus rex* in the same way that they are more willing to buy a Picasso, Renoir or Monet than any other twentieth-century artist.

In the future, though, he sees this changing. When my nephew's generation grow up, they'll probably look beyond the familiar *Tyrannosaurus rex*, *Stegosaurus*, *Triceratops*. Hyslop remembers telling his own nephew about a *T. rex* tooth he was selling. 'I thought he'd find it interesting, but straight off the top he said, "Spinosaurus teeth are larger." He wasn't impressed by *T. rex*.'

Not everyone is comfortable with the dinosaur as a status symbol. In 2018 the French auction house Aguttes advertised the sale of a large, as yet unclassified dinosaur. The Society of Vertebrate Paleontology (SVP) in Bethesda, Maryland, which represents more than 2,200 international palaeontologists, wrote to Aguttes, urging the auction house to cancel the sale. 'Fossil specimens that are sold into private hands are lost to science,' the letter states.[18]

And even if a private collector allows a scientist access to their fossil, there's not much the scientist can do with the material. 'Fossils have to be publicly available to study, otherwise you can't publish papers on them,' Maria McNamara of University College Cork explained to me. The results need to be available to be replicated and checked, and with a fossil in a private collection that can't be guaranteed. 'I do think about all the scientific wonders that are potentially locked away in Hollywood starlets' houses,' she said. 'With chemistry you go out and you synthesise your chemicals and study how they form.

With biology you take your cells and you study how the proteins move in and out. With palaeontology we can't just readily get our data. We are dependent on a limited number of very good fossil localities.' Although the SVP doesn't collect data on the number of dinosaur fossils sold at auction, Paul David Polly, its former president, says that 'very-high-priced auctions are becoming much more common.'[19] Museums don't have the budgets to compete.

But there are dissenting voices. 'On that question I part with my colleagues,' Carpenter told me, when I asked if he agreed with the SVP's campaign against private sales. 'If we look at the history of vertebrate palaeontology, it begins with people buying and selling. Mary Anning sold a lot of her stuff to the Natural History Museum. In this country Marsh and Cope used to buy specimens.' Vinther agrees. 'The thing is that if we said it was completely illegal to collect fossils, then nothing would ever be found. We [academics] don't have the time to go out into the field constantly, so we are actually dependent on people finding these things.' He argues that scientists need to develop better relationships with the commercial fossil-hunters instead of stigmatising them.

The Aguttes dinosaur was eventually sold at auction in Paris for over 2 million euros, to an unnamed bidder. It was reported that the owner would like to see it in a French museum, but at present no more is known.

While these are huge amounts of money, Hyslop believes that the market is reaching a cap in some areas. 'For example, at the moment it isn't commercially viable to excavate a *Triceratops*,' he told me. All fossils are rare, and dinosaurs especially so, but within those parameters *Triceratops* are relatively common. 'I describe them as the cows of the Cretaceous,' Hyslop said. 'If someone says, "I found a dinosaur bone," odds on it will be a

Triceratops bone.' This is partly because there were at one point a lot of *Triceratops* roaming about, and partly because they are relatively late dinosaur fossils, so the rocks they're found in are relatively young and have had less time to be deformed and destroyed. 'There seems to be a ceiling of about half a million for a complete *Triceratops*, and the skull will make about three or four hundred thousand if it's a really good one, but you will spend 10 per cent of your time getting the skull out and 80 or 90 per cent on the rest of the animal.' Commercially, it makes sense for a dinosaur-hunter to pick up the skull and leave the rest of the body behind. 'Which is heart-breaking, isn't it?'

Indeed, in the context of the art market even the $8.3 million paid for Sue is relatively modest. 'During the bone wars the price for a top dinosaur specimen was pretty much on a par with the top paintings in the world,' Hyslop said. As of 2020 the world's most expensive painting sold at auction is Leonardo da Vinci's *Salvator Mundi*, which went for $450 million in 2017. 'For that price I could put together the best collection of dinosaurs anywhere on earth, trumping any natural history museum in the world.'

＊

But the thing we think about most when we think about dinosaurs is that they ruled pretty much everywhere on Earth until, one day, they didn't. They have always been inherently tragic figures because it's difficult to think about them without thinking about the fact that they are not here any more.

'The thing that killed the dinosaurs was a meteor. It exploded, and there was gas and molten rock. And the volcanoes. There was smoke, and the sun got blocked out and that made the plants die and then the herbivores couldn't eat. So they died and then the carnivores couldn't eat, and that's why they all died.'

A meteor and volcanoes. My younger nephew has a decent grasp of what many scientists now speculate may have happened during the K-Pg extinction event (K for Cretaceous – C is already used to denote Cambrian – and Pg for Palaeogene, the two time periods on either side of the extinction event). To his description we might add that the meteor or giant asteroid landed in Mexico and created the Chicxulub crater. And that in addition to general increased volcanism the dinosaurs also had to contend with the Deccan Traps in India. There, over hundreds of thousands of years, molten lava spilled out of an area the size of France, Germany and Spain combined. In places the lava was an incredible 150 m thick. It's likely that all this volcanic action resulted in acid rain and contributed to phases of both global cooling and global warming that happened during the late Cretaceous, further destabilising Earth's ecosystems and stressing animal populations.

There's also evidence that in some areas (western North America, for example) dinosaurs were already in decline at the time of the meteor impact – scientists point to a lack of diversity among dinosaur groups at the end of the Cretaceous, possibly the result of habitat change caused by a fall in sea-levels. The less diverse a group of animals, the more vulnerable they are to extinction. As Naish and Barrett write: 'The set of dinosaur species present [...] in western North America therefore looks like an "extinction-prone" community.'[20]

Whatever the exact cause, it is estimated that around 75 per cent of all marine, and 50 per cent of all terrestrial species vanished: not just the non-avian dinosaurs but also certain mammals, lizards, insects, pterosaurs and plants, and in the ocean plesiosaurs, mosasaurs, ammonites and many fish and sharks.[21] After the K-Pg event, the new world would have been a quieter, emptier place – the oceans largely silent, the land covered with drifting ash and charred tree stumps, first a place of extreme cold and then, as global warming kicked in, hellish heat.

Now more than ever, as catastrophes that took place in the past, in deep time, form an irresistible backdrop onto which to project our own fears of climate change and contemporary apocalypse, we look at the dinosaurs – a fellow worldwide apex predator – and shudder. Dinosaurs were big, lethally well-adapted creatures, and in the end that didn't help them.

'Were they not rulers of their empire, just as we are of ours?' Jan Zalasiewicz has written in *The Earth After Us*. 'Might we not hope for similar awe and reverence from our future excavators?' He imagines palaeontologists of the future, extraterrestrial perhaps, digging up the fossilised remains of our bodies.

> [But] perhaps their focus will be on what, among all the diverse inhabitants of this planet, is most important in preserving the living tapestry. They may well regard the myriad of tiny invertebrates, or the bacteria, of the world as much more important to that (in planetary terms) rare phenomenon, a stable, functional, complex ecosystem [...] Take away the top predator dinosaurs, and the Jurassic ecosystems would have been a little different, to be sure, but no less functional. Take away humans, and the present world will also function quite happily, as it did two hundred million years before our species appeared. Take away worms and insects and things would start seriously to fall apart. Take away bacteria and their yet more ancient cousins, the archea, and the viruses too, and the world would die.[22]

*

Near the town of Moab, Carpenter had told me, there are dino-saur trackways preserved on an old copper-mining trail. The day was hot again, the landscape shadeless. Walking uphill, the flat sandstone rocks underfoot are a mottled, dull, rusty red, almost purple in places. Their uneven surface is inscribed with looping marks – ripples that formed 150 million years ago, when the rock was a sandbar in a shallow Jurassic river, one of the many that criss-crossed the landscape we would come to call the Morrison Formation.

On either side of the path were yellow rocks cropping out of yellow-grey soil. Clumps of blue desert grass topped with pale golden feathery tops. Stunted cedar trees with bright green leaves and tiny, dusty blue-grey cones. A black bird – raven, was it? – winked in and out of sight against the blue sky.

One hundred and fifty million years ago three dinosaurs passed this way: a large sauropod and two theropods, one injured and limping. They left their footprints in the sandbar, and over time these footprints were filled with sediment and buried and turned, eventually, through the processes of deep time, into rock. The sauropod's footprints were big, blunted shapes – you'd have to know what you were looking at to distinguish them from a random depression in the rock. The theropod footprints were smaller and more recognisable as animal markings, something like a cross between an enormous dog and a bird, each print with a clearly distinguishable pad and long, lozenge-shaped toe impressions.

I sat down and drank warm water from my metal canteen. In the distance the flat-topped buttes looked like clouds. Over-head the clouds themselves receded backwards in discreet, parallel lines like the layered scenery in a theatre. As I watched, the bird came back into sight, landed a couple of metres away on a limestone boulder and cocked its head, eye bright with some avian intelligence.

Since the 1990s, new fossil sites – many in China, which has become, as the US once was, a rich new frontier for dinosaur-hunters – have yielded exceptional material showing how, around 160 million years ago small, feathered, predatory dinosaurs evolved into birds. Today most palaeontologists also agree that modern birds are not merely related to or descended from dinosaurs – they actually *are* dinosaurs, classified as a sub-group of the theropods (a group containing, among others, *Tyrannosaurus rex* and *Allosaurus*).[23]

I crouched down with my notebook to sketch the shape of the tracks in the rock. Kneeling there in the Utah desert, it seemed to me that, more than fleshless bones in a museum – which must emphasise deadness, gone-ness – it was these marks that finally brought me closer to the animal itself. I thought about something David Norman, curator of Cambridge's Sedgwick Museum, had said about trackways during a talk I'd been to at the Geological Society: 'A dinosaur went past and left that behind. In a way it's moving to think that if you touch that specimen you're as near as you're ever going to get to a dinosaur.'

Talking about the non-avian dinosaurs, it's easy to speak about everything except the reality of the beasts themselves: the hype, the myth, the market-place, the commercialisation, what dinosaurs tell us about ourselves and our cultural moment. Bombarded with plastic figurines and dinosaur memes, familiarity stops me really *seeing* a dinosaur. Imagination slides away. Dinosaurs. Sure.

But then something snags your attention. Touching the edges of one of the theropod's lozenge-shaped toes, 150 million years almost dissolve in the hot desert air. There really were dinosaurs here. For a moment that fact becomes amazing once again. *There were dinosaurs here.*

12

COLOURING DEEP TIME

The rail – a small, coot-like bird – had died one day during the Eocene Epoch, on the piece of land we now call Denmark. Around 55 million years later a Danish PhD student placed the fossilised remains of the bird underneath a microscope. Not that this was much consolation to the rail, but it was about to make a major contribution to palaeontology.

Jakob Vinther is now a lecturer in palaeobiology at the University of Bristol. He talks very quickly in slightly accented English, waving his hands, often ending a sentence with *bammm*, or *blahh* or *vroosh*. As a child in Denmark he was obsessed with the natural world, keeping his own collection of utility plants – fruits, vegetables, tobacco and even hemp (discovered growing in an abandoned allotment) – and crowding his bedroom with aquariums filled with lizards and snakes. The whooshing of the air pumps used to keep him awake.

The story of his work with the rail begins in fact with a much older creature: a fossilised 200-million-year-old squid relative.[1] Since the nineteenth century fossil-hunters including Mary Anning have known that prehistoric ink is sometimes preserved in the sacs of these fossils. There was ink in the specimen Vinther was studying in the lab at Yale where he completed his PhD; it proved a 'Eureka!' moment. 'I was looking at this ink and thinking, "What the fuck?", and realising that it's identical to the ink in a living squid, which means it's composed of melanin. That's the same pigment that we have and dinosaurs have.'

By studying the pigments it might be possible to do something long dismissed as impossible – to use the evidence from a fossil to work out the colour of a long-dead creature.[2]

Pigments produce colour by the selective absorption of certain wavelengths of visible light. Typical pigments include melanins (reds, yellows, browns and blacks), carotenoids (bright reds and yellows) and porphyrins (greens, reds and blues). Other colours are produced by light-scattering nanostructures. These are known as structural colours because, instead of using chemistry, they create colour through the use of microscopically structured surfaces fine enough to interfere with visible light. Think of the brilliantly coloured feathers of some tropical birds or the buttery sheen on a tulip leaf. Many plants and animals use a combination of pigment and structural colour. Peacock tail feathers have a brown pigment, for instance, but microscopic surface structures mean that they also reflect blue, turquoise and green light, and give them their famous iridescent sheen. In humans, melanin controls our hair and eye colour, among other things. It is produced and stored in tiny cellular bags called melanosomes, which come in two forms. A sausage-shaped type produces black shades, while a round-shaped variety creates rusty reddish hues. If you have red hair, you have round melanosomes; if you have black hair, you have sausage-shaped ones; if you have brown or grey hair, you probably have a combination of the two shapes and some pigment absence.

In exceptional circumstances skin and feathers will fossilise, though the colour of the fossil (typically black or brown) is the result of the processes of fossilisation and not a guide to the colour of the living creature. But if pigments had been preserved in the ink sacs, Vinther reasoned, then melanin – or the melanin-bearing melanosomes – might also be found in fossilised skin and feathers. Now he just needed one of those incredibly rare soft-tissue fossils to test his theory.

It wasn't easy. Given their scarcity, curators are reluctant to allow just any unknown PhD candidate to slice into such fossils, producing a sample small enough to fit into the powerful electron microscope that Vinther would need to use. Eventually he persuaded the curator of vertebrate fossils at the Geological Museum in Copenhagen to cut down a typewriter-sized block of limestone containing the skull of the 55-million-year-old rail.

For something so dead, the bird, when Vinther showed it to me, still looked surprisingly alert, head cocked, watching for some tasty Lower Eocene grub. What made it so lifelike were the feathers. It has been preserved with a dark halo of feather impressions and two stains where the eyes use to be. It didn't look like a skeleton so much as an animal that had been preserved like a dried flower pressed between the pages of a book. When Vinther first took possession of the specimen, he fitted it into his microscope and began scanning. 'I was sat there zooming in with the microscope, looking for the melanosomes and suddenly I was like, blimey, they're there! We can put colours in fossil dinosaurs.'

At first his supervisor Derek Briggs was sceptical. The structures Vinther described were already well known and classified – as bacteria.[3] 'They are the same size and shape as bacteria and they're found on these rotten carcasses where you would expect to find decay bacteria,' another scientist told me. 'It all seemed very plausible.' Seeking further evidence, Vinther and Briggs looked at a fossil feather from the Cretaceous period with distinct black and white colour bands. Where the feather was black there were sausage-shaped melanosomes; where it was white there were no melanosomes (white indicates an absence of pigment). Had the melanosomes been bacteria, they should have been seen on both the black and the white parts of the feather. 'It was something of a lucky strike,' Vinther says now. 'I think it's just because I came with relatively new eyes as a first-year PhD student, and came from a relatively safe point – I had no reputation to lose.'

Vinther published his initial findings in 2008.[4] Now the race was on to produce the first coloured dinosaur, using the shape of the melanosomes to deduce hue and pattern. In 2010 two teams at the University of Bristol, one led by Vinther and the other by Michael Benton, published within days of each other. They showed, respectively, that the birdlike *Anchiornis huxleyi* was crowned with a red crest and that the feathered dinosaur *Sinosauropteryx prima* had a reddish-brown striped tail.[5] Since then, further studies have built on Vinther's original hypothesis, including work by Johan Lindgren from Lund University, who used Time-of-Flight Secondary Ion Mass Spectrometry (ToF-SIMS) – a highly sensitive analytical technique that describes the chemical composition and distribution of a sample surface – to analyse the composition of various fossils.[6] He found direct evidence of the chemical signature of melanin pigments. (This is as opposed to the indirect evidence of the shapes that correspond to the shapes of melanosomes, and from which information about colour can be inferred.)

Finding out the colour of an extinct creature feels psychologically important. A little piece of the past becomes clearer, pulled out of the blurred confusion of deep time and into focus.

We are a visual species and give primacy to the sense of sight. It makes things 'real' for us. 'Inevitably,' wrote the archaeologist and nature writer Jacquetta Hawkes in 1950, 'the senses are demanding of the imagination. Has it colour? But the imagination, with no more to work on than a poor tangled skeleton lying among delicate tissues of feather impressions in the grey monochrome of the limestone, admits itself defeated.'[7]

So far only a handful of dinosaurs, insects and reptiles have been studied. But, as Lindgren says, we're only just scratching on the surface. Previously, an illustrator drawing, say, a *Tyrannosaurus rex* would use informed guesswork. Should he evoke the sorts of earth tones associated with many modern reptiles and amphibians? Or look to the bright, flamboyant plumages of some modern birds – the only dinosaurs to survive into the twenty-first century? Now we have the beginnings of a method that will turn the black-and-white prehistoric world into glorious technicolour.

*

Like Vinther, Maria McNamara is a palaeobiologist. Fifteen years ago she would probably have called herself a palaeontologist, but simple palaeontology is these days considered quite old-fashioned, bringing to mind elderly men in tweeds debating the finer points of classification, sunburnt men in safari jackets stumbling back from the desert clutching bones, 'whereas an awful lot of palaeontologists are much more interested nowadays in the biology of ancient animals. Not so much what species it was but, how it moved, what it ate, where it lived, what colour it was,' said McMamara.

Now in her late thirties – she was a post-doctoral researcher at Yale at the same time Vinther was completing his PhD

– McNamara often spent her summer holidays as a child roaming the fields and woods of north Tipperary, where her grandmother lived. 'She sent us out and said, don't come back until you've found three different grasshoppers or five different types of pink flower or whatever it was. We'd always be outside in nature, and I think that was very fundamental – I was always really interested in looking at plants and animals and thinking about how the planet works. And really as a scientist all you're doing is just looking at things – taking the time to properly look.'

Today she works at University College Cork and has pioneered work on the preservation of non-melanin coloration methods, including the first systematic investigations of the fossil record of structural colour. In 2011 she used incredibly powerful microscopes to study the shape of the structures that produce structural colour in fossilised beetles.[8] In 2016 she published the first paper showing evidence of the preservation of carotenoid pigments, reconstructing a 10-million-year-old green-and-brown patterned fossil snake from north-eastern Spain.[9] 'We did it because we wanted to show that it was possible for non-melanin coloration methods to preserve in fossils. Up until then people had only been looking at melanin. And still everyone is only looking at melanin. But that fossil showed, look, you can preserve the carotenoids.' That evidence that both carotenoids and structural colours may in some instances be preserved can, she says, 'dramatically modify the colour that you'd predict for a creature based on the melanin alone'.

While she was at Yale, McNamara decided that she wanted to know what happened to differently coloured feathers during the fossilisation process.[10] As she didn't have hundreds of thousands of years to wait for a fossil to form, she decided to speed the process up and make her own.

You can use heat and pressure to approximate the effects

of fossilisation. In the basement of her department, McNamara borrowed a colleague's rig – in effect, a very powerful oven where you could control both temperature and pressure. Another colleague pointed out that the pressures generated by the machine were such that, should anything break, the oven would shoot up through the two floors above. McNamara was relocated to a concrete bunker on the edge of the campus and the machine placed behind a heavy steel door.

Fifteen feathers were chosen. All contained melanosomes, but some also contained other pigments and structural colour mechanisms and were yellow, bright blue, red, orange or green. Each was wrapped in tin foil and cooked for twenty-four hours at 200°C and 250 bars of pressure. At the end of the experiment McNamara found that only the melanosomes – controlling browns, blacks, rusty reds etc. – had survived the fossilisation process. 'So if you interpret colour based on melanosomes alone you're just fooling yourself. You've lost evidence for other pigments and structural colour mechanisms in the feathers.' The second point her experiment showed was that the geometry of the melanosomes changed, suggesting that it's not possible to predict absolutely precisely the hue of the extinct creature – though the general shapes, whether it's a sausage shape (black) or the round shape (rusty red), were preserved. Vinther counters that he had already accounted for the fact of shrinkage when he published his papers. The debate continues.

For McNamara, a big concern is that melanin itself is not yet properly understood. 'We need to learn more about melanin in modern animals before we go near fossils.' Much more than just colour, melanin can provide, for instance, UV protection and mechanical strength. This is the reason some birds have dark wing tips: the melanin makes the surface of these vulnerable outer feathers stronger and more resistant to abrasion. McNamara is particularly interested in the fact that melanin exists not just in hair and skin but also in internal organs. 'We're going to try and find out what is controlling melanin evolution. We've always thought, oh, it's for colour, oh, it's for sexual selection and camouflage, but if it's in all these internal organs then maybe it evolved for a completely different purpose. And you don't want to be remembered for getting it wrong. That's absolutely the worst thing that could happen as a scientist.'

Lindgren agrees. My feeling is that as palaeontologists we have a tendency to oversimplify, to think A must point to B. But if you were to ask a biologist, we know that in the modern world there is never ever a single factor that leads to one thing.'

Other scientists have argued that one simply can't extrapolate the colour of a bird or dinosaur from a single isolated feather – or from the minute samples that are used in the ToF-SIMS technique. Imagine trying to determine the coloration of a modern-day peacock from pigments taken from just a few spots, said Mary Schweitzer of North Carolina State University in an article in the *Proceedings of the National Academy of Sciences* (PNAS).[11] There are concerns that the new field is developing too quickly, that claims are being made which, Lindgren suggests, 'state more than we actually know'. Contentious issues are fiercely debated, and several of the people I speak to have somewhat tense things to say 'off the record'.

Part of the reason for all this tension is that the emerging field of palaeocolour is what scientists call 'high-impact', meaning that it's the stuff that wins grants and gets a *Nature* cover and press attention. Success in this area could transform a scientist's career.

But for Vinther, the continual arguments about palaeocolour

go too far. Every time he publishes a new paper, he wonders what people are going to say, what he will have to respond to. 'You spend a lot of time in competition with these people and you think, oh god, it would be nice just to do something that people care about but where they don't get so antagonistic.'

*

Beyond showing an illustrator the correct shade of Pantone to use, what can an extinct animal's colour tell us? The answer is quite a lot.

Bones fossilise: behaviour, how animals interact with one another and their environment can't. 'When we look at the animals and plants we see in the world around us, we see striking colours and colour patterns,' said McNamara. 'Animals use colour for camouflage, for avoiding predators, for mating signals and also for signalling within their social group. So evidence of colour in animals has the potential to tell us about this very enigmatic aspect of the biology of ancient organisms.'

It can bring new insights into the daily lives of long-dead creatures. For instance, it had long been presumed that the small, four-winged *Microraptor* was nocturnal, based on large eye sockets. Then Vinther, Quanguo Li from the Beijing Museum of Natural History and colleagues discovered that the dinosaur possessed iridescent plumage. This would have made no sense if the dinosaur were active only at night.[12]

In the future scientists may be able to chart the progress of colour through deep time, answering questions such as, what is it that drives colour evolution? Is it natural selection – the desire to hide yourself – or sexual selection – the desire to advertise? McNamara's beetles, for example, probably used their shiny structural colour in mating displays, as some modern beetles do today. A beautifully coloured male beetle is signalling that he would make a suitable mate: after the basic business of surviving is done, he still has resources to throw into making beautiful, energy-intensive structures.

'But was there ever a time when competition pressures were less and sexual selection wasn't happening?' McNamara wonders. 'What would colour patterns look like if they weren't controlled by those factors? Would you just get really crazy patterns? Or no patterns at all?' Looking at the way modern animals behave can give us clues, but we can't assume that the world has always been the way we see it today.

Colour can also tell us about the environment an animal lived in. Scientists try to gather clues about this by looking at other fossil animals and plants found near by – but this technique falls down if the animal's body has been transported by, for example, a river, away from the place where it lived. Vinther studied the fossil of a small, plant-eating dinosaur called *Psittacosaurus*, a relative of *Triceratops*, and concluded that it had a dark back and pale belly – a colour arrangement known as counter-shading.[13] Common among modern animals ranging from whales to deer, it is used by both predators and prey to blend in with their surroundings. (Parts normally in shadow are light; parts normally exposed to the sky are dark.) The amount and distribution of the light and dark areas typically correspond to the quality of light in different habitats. The counter-shading on the *Psittacosaurus* suggests that it lived somewhere with diffuse light, such as a canopy forest.

The adaptation of camouflage – whether to avoid being eaten or to get closer to your dinner – is part of a general pattern of escalation, an arms race that has been going on since the 'Cambrian'. It tells us not just about the animal's environment but also about the other creatures it shared that environment with. 'When animals attack each other, they then have to adapt. Like the Red Queen in *Alice Through the Looking Glass*, you have to keep running to stay in the same place,' Vinther said. Think about really big modern herbivores such as elephants and rhinoceroses. The reason they are not camouflaged is because they are too big to be eaten, and their 'prey', grasses and trees, can't run away. Then think about an *Ankylosaurus*. This herbivore was heavily armoured and grew up to 8 m in length, weighing as much as 8 tons, a little more than an African elephant, the largest land creature now living. Today, such a creature wouldn't need to be camouflaged. In the Jurassic, it had counter-shading.

'Something made this poor *Ankylosaurus*, armoured to its teeth, also be in need of camouflage,' Vinther said. 'And that tells us that there were some very, very scary predators around back then. Jurassic Park was real.'

*

When Vinther was working on the coloration of the *Psittacosaurus* fossil, he enlisted the help of a palaeoartist called Robert Nicholls. Worldwide, only a handful people make a living from palaeoart, and Nicholls is one of them. I went to visit him in his studio at the top of a small terraced house in a suburb of Bristol. The walls of his studio are decorated with framed *Nature* covers featuring his illustrations and, somewhat incongruously, a reproduction of J. W. Waterhouse's *The Lady of Shalott*. 'My god I love that painting,' Nicholls said. 'There's a real economy of mark-making. From a distance it's utterly convincing, but when you get up close you can see they've just whipped it on there.'

I sat watching while he modelled the head of a *Tyrannosaurus rex* on his computer screen. From the small park across the street came the sounds of children playing. Downstairs his wife and young daughters were having an early tea. As Nicholls worked, the snarling dinosaur head span round and round on the screen, vivid slashes of red across each cheek like war paint. Palaeoartists are commissioned by museums, scientists and publishers to create paintings and models of long-dead prehistoric creatures. For the most committed, their work goes far beyond simple illustration. 'The reconstruction process is what defines palaeoart, looking at the fossil material and building things from the inside out,' Nicholls explained. 'It's not something every artist can do.' Reconstruction involves thinking, in the same way that a palaeontologist would, about how the animal is put together, from the bones, through the muscle to the skin, basing your speculations on the available fossil evidence.

'In my industry there's a thing called shrink-wrapping,' he told me. 'It refers to people who don't put much tissue on their dinosaurs.' A shrink-wrapped *Tyrannosaurus rex* would be one where the artist has, in effect, simply wrapped a skin straight around a skeleton, so that you can see the shape of the skull, the shoulder blade poking out and each of the vertebrae on the back. 'And of course that's not what a real animal looks like. With the *Tyrannosaur*, for example, there's an opening in front of the eye that's often illustrated as a depression, but it would have been full of soft tissue, so if anything it would have bulged out.' Nicholls pointed to the screen. 'When it's finished, this *Tyrannosaur* should hopefully look utterly real, as though you could touch it.' His style is what we might call 'photorealist', had cameras existed 66 million years ago. 'I suppose what I really want is to be able to time-travel and see dinosaurs,' he said. 'Anything very dead that's turned to stone,

I want to be able to see it, and it frustrates me that I can't. So being able to do this and make it look utterly convincing is what excites me.'

Most professional palaeoartists are palaeontologists by training, but Nicholls has no formal scientific background. He struggled with dyslexia at school, gaining a double E in combined science – but in order to make his reconstructions as accurate as possible he reads as much palaeontology as he can, listens to audio books and attends scientific conferences. To gain insight into the biology of these long-dead creatures, he also studies the physiology of modern animals, especially birds. Not long before we met he had attended the dissection of an ostrich at the Royal Veterinary College.

When Vinther contacted him in 2014 to discuss the *Psittacosaurus* model, Vinther explained that he had come across an extremely rare fossil where much of the creature's soft tissue, including evidence of melanosomes, had been preserved. Using this material would allow them to create the most accurate – or, strictly speaking, credible – dinosaur model to date. Nicholls needed no persuading.

It took almost four months of work, beginning with several days at the Senckenberg Museum in Frankfurt to photograph and measure the fossil. 'I've measured every single bone and the curvature of the ribs to get the shape of the torso right, the proportion of the limbs,' Nicholls said. In the early stages of the build, he worked on paper, drawing muscles and soft tissue on top of the skeleton. 'You need to work to forget any preconceived ideas in your head and just follow the fossil evidence. You know you're going somewhere right if the drawing surprises you.' With the *Psittacosaurus*, this method worked almost immediately. Growing up, the reconstructions of *Psittacosaurus* that Nicholls had seen were all quadrupeds, so that's what he'd expected to model. But when he began reconstructing the animal's anatomy, the proportions of the limbs and a rigid backbone suggested instead that it was probably bipedal.

Next, Nicholls sculpted a clay model, which he used to build a liquid silicon mould and finally a fibreglass model. The painting process typically takes around a week, but in this instance there was so much information from the soft tissue and Vinther's new work on coloration that it took almost a month.

At the end, after the insertion of two yellow glass eyes, Nicholls was able to step back and see a creature dead for 101 million years staring back at him. *Psittacosauruses* have been drawn and sculpted before, but never with this degree of care, attention and information. So far, Nicholls has produced two models – Podge (living in Vinther's office) and Stanley, who sits on a workbench in the corner of Nicholls's studio. About the size of a labrador, Stanley has a distinctive parrot-billed beak and a wide head with horn-like structures of either side of his face, something like Princess Leia buns. ('For this species of dinosaur it seems that a wide head was a really sexy feature,' says Nicholls.) A dark-brown and orange mottled back becomes progressively paler, down to a creamy underbelly. The look in his yellow eyes is engaging, friendly even.

'There was a point when I was sculpting it and I thought this looks kind of cute. Is this what I want? Perhaps I should make it look more ugly?' Nicholls said. 'And then I thought, that's just pandering to what a lot of people expect to see. They think dinosaurs are movie monsters, but they're not, they're real animals. There are plenty of animals today that look cute, so why shouldn't a dinosaur be cute?'

When Vinther's paper on *Psittacosaurus* counter-shading was published, it drew directly on what they'd learned during the creation of the model. Nicholls was one of the named authors.

Projects such as the *Psittacosaurus* are relatively rare. Often clients can't afford to pay a palaeoartist to spend the time properly researching and building their models and pictures from the skeleton upwards. 'We could go to a bookshop and I could take a book off at random and see straight away whether it's been done by a generalist or a specialist illustrator. A generalist will almost always copy other artists because they don't know how to do the reconstruction process,' Nicholls told me.

Artists who lack the resources or time or knowledge or inclination often resort to copying images that already exist. 'That's why there are so many stereotypes or memes in paleo-art,' Nicholls said. He fears that around 90 per cent of the art that gets published is 'regurgitated nonsense'. In one particular piece, he remembers, the artist had copied from others so slavishly that each different part of the illustration was in the style of a different copied artist, giving a peculiar collage-like effect.

Nicholls showed me a piece of artwork he has just completed for the cover of the new edition of the Natural History Museum's book *Dinosaurs: How They Lived and Evolved*.[14] The first edition had been heavily criticised by what Nicholls refers to as 'the hardcore palaeoart community' because the cover – showing a roaring theropod – was anatomically incorrect and a huge cliché. When Nicholls got the commission, he decided to do something totally different.

The enigmatic painting that he produced shows *Tianyulong*, a recently discovered dinosaur from China. Drawn against a black background, this dinosaur has yellow, owl-like eyes surrounded by red folds of skin. Wispy ginger hair, standing up on end as though charged with static, covers the head and body. Large claws clasp at a branch of green foliage, pulling it towards a beak-like mouth. This is a picture of a dinosaur that celebrates its animalness, and hence familiarity – it is not a movie monster – and which at the same time jolts us out of our lazy expectations and lets us really *see* a dinosaur again, recapturing the sense of strangeness and wonder that the Victorians must have felt when they first glimpsed these long-dead reptiles. It was a brave move by the publishers: an attempt to redefine what people think of as a dinosaur. Not some *Jurassic Park* scaly beast but this gently weird, plant-eating creature. 'Hopefully it won't negatively affect sales, otherwise we'll be condemned to another decade of roaring dinosaurs on the cover of books,' Nicholls said.

On his computer he brings up another image. 'This is a photo composite showing a herd of herbivores crossing a river, and so far it has taken about 100 hours of painting.' To produce such a picture he models each creature individually and then stitches together a background from about 100 photographs to create his version of the Triassic. 'When I was younger, my pictures were all about the gore. This scene would have bored and horrified me, but now it seems really exciting. I like taking a familiar occurrence that we've all seen in pictures and films, such as a herd of wildebeest crossing a river, and making it happen in an unfamiliar environment with unfamiliar animals.'

For Nicholls, the whole industry of palaeoart is growing up and changing focus. 'There's a few of us that take this seriously, and we're trying to produce a more diverse, complex, natural type of reconstruction.' In the future the harnessing of cutting-edge scientific techniques, especially in the field of colour reconstruction, is likely to play an increasingly prominent role. 'What I really like about colour reconstruction work is that you're the one defining what an animal looks like, down to the colour pattern, for the first time,' Nicholls said, smiling. 'Being able to show people something that no one has ever seen before – that is the best.'

MANUFACTURED LANDSCAPES

13

URBAN GEOLOGY

Wine-red Griotte d'Italie against moss-green Connemara marble. Dark green, smoky Verde Alpi serpentinite. Blue Belge, Belge Noir, pale pink Nembro Rosato the colour of an old-fashioned tea rose. Pillars of buttery gold Siena Breccia, dove-grey Repen Zola, translucent, orange-tinged English alabaster.[1]

I'd gone, one day, to the fever dream of the Brompton Oratory in South Kensington, shining piles of marble and alabaster turned into columns, panels, altars, fonts and sepulchres. Stilled swirls, branching veins, jagged stripes and smoky clouds of colour. Marble is a happenstance of physics and chemistry – the effect of intense heat and pressure on sedimentary limestone, the ingress of mineral-rich waters. The French philosopher and collector of stones Roger Caillois once argued that we find the marble beautiful because the marble itself, far older than human civilisation, has taught us what is beautiful. The stones themselves have shaped and guided human aesthetics.[2]

In the marble slabs and columns we can trace journeys through tens of thousands of years. A brightly coloured vein is where hot hydrothermal waters, filled with exotic minerals, pulsed through a crack in the original limestone. A jumble of crystals in the rock's matrix are testament to millennia spent slowly forming, transforming, in the heat and pressure of some underground darkness. The signature of deep time itself written in the rocks.

*

Since researching this book, I've become addicted to these glimpses of deep time in the modern city. I stare at the stone facing on empty walls, interrogate pavements and the spaces between shop windows, the lintels of doorways, the sides of bridges. Looking for deep time, the fabric of the city comes alive in unexpected ways. In grey sandstone paving stones a series of fine bands of curving lines are the fossilised ripples from the current of a river that, 300 million years ago, flowed through what is now the Pennines. Sunlight glancing off the silvery grey granite facing the entrance to Topshop at Oxford Circus catches scales of mica, sets the façade shimmering, trembling, as though about to collapse. On the concourse of Paddington Station you can trace a skeletal, segmented cone-like shape: the 450-million-year-old fossil of a snail shell embedded in the limestone floor tiles.

In the 1970s a geologist at University College London called Eric Robinson began exploring the idea that geology could be taught not just on expensive, far-flung field trips but also through the buildings and streets of our cities. An early proponent of the idea that there is historical and cultural value in the serious study of building materials, he pioneered the work known today as 'urban' or 'street' geology. For the urban geologist, the city becomes a sort of enormous, jumbled specimen cabinet. A site of geological wonders. A democratiser and enthusiastic educationalist, Robinson published a series of geo-walks around London for students and members of the public, many of which can still be followed today.

One day in April I went to Waterloo Station to meet Ruth Siddall. Also of University College London and a former colleague of Robinson's, she has done much to expand on his work, leading and writing walks, as well as setting up a catalogue of London building stones, a web site and an app – London Pavement Geology – that allows members of the public to view and add to a database of the buildings stones of the city.

Many of the famous landmarks that make up what we might think of as 'iconic' London, Siddall told me, came about because of conditions in the Jurassic seas around 145 million years ago, when a type of limestone we call Portland Stone was forming in what is now the east of Dorset. On the station wall, Siddall showed me how the seemingly smooth surface of the rock is actually formed from millions of ooliths – tiny spheres of calcium carbonate moulded by tidal action – and broken-shell sand with fragments of grey oyster shell still visible. First transported to London by sea, Portland Stone was used by the architect Inigo Jones to build the Banqueting Hall in Whitehall for James I, and then extensively by Christopher Wren and Nicholas Hawksmoor in the construction of their churches. It's what University College London and the British Museum are built from. Its popularity derives from its accessibility, plentiful supply, weather-resistance and the fact that it is what is known in the trade as a freestone, meaning that it can be cut in any direction.

As we walked towards the Thames, Siddall took pictures and made notes for the catalogue of London stones that she is compiling in collaboration with the Geologists' Association – an organisation founded in 1858 to cater for the needs of amateur geologists as opposed to the 'professionals' at the Geological

Society of London.[3] Two teenagers stared curiously as we stopped to photograph some granite at the base of an office building. 'Doing this you have to adopt a no-shame approach to looking like a complete geek in public,' Siddall said.

For much of history and in most places people have tended to build their towns and cities using whichever rocks were closest to hand. This was never possible in London, a city built on sliding clays and soft chalk. Bricks were once made locally – those rich honeycomb-yellow Georgian bricks formed from London Clay – but chalk is too soft to make a satisfactory building material. Instead the stones of London, Siddall told me, 'come from all over the world'. On the South Bank, outside the Royal Festival Hall, we saw a 2-billion-year-old glassy black gabbro from Pretoria, forming the plinth for the giant bust of Nelson Mandela. Inside, visitors milled around, banners advertised a performance by the Simón Bolívar National Youth Choir of Venezuela, and we looked at slabs of Derbyshire Fossil Limestone – a remarkable purplish stone filled with galaxy-like swirls of paler shapes: discs, rectangles, things that look a little like tuning forks and other things that look like chrysalises. They are, in fact, the fossilised remains of a grove of crinoids or sea lilies, creatures from Carboniferous times. 'This is the best example of a crinoidal limestone that I've seen anywhere in the world,' Siddall said. (Several weeks later, joining Siddall on a walk she'd organised for the public, I found myself facing another slab of limestone, this time on Charlotte Street in Fitz-rovia. Studying the tea-stain-coloured limestone wall of the Muslim World League building, I turned to the tall man standing next to me and asked if he could see the ammonites we'd been told to look for. We squinted at the rock for a bit. He was a geophysicist, not a palaeontologist, he said, a bit disconsolately, but – brightening up – he did come from the Cotswolds, which was where this limestone had been quarried. By the time we reached the BBC building on Portland Place – fossiliferous Portland Stone – we'd decided to go for a drink. Two years later we were married.)

Siddall's favourite building stones are granites. 'They're sort of like ice-cream: the same basic recipe but a whole world of flavours.' For granite, an igneous rock, the ingredients in the recipe are feldspar, quartz and mica. Most London kerb-stones are made of granite, but when polished as a decorative stone the colours range from a kind of oatmeal – the Cornish granite at the base of Waterloo Station – to the dark red and salmon pink – like a dish in aspic in a 1970s cookbook – of the Peterhead granite from Aberdeenshire that forms the pedestal for the statue of Queen Victoria at the northern end of Black-friars Bridge. For a moment Siddall and I stopped to watch a boat filled with tourists moving down the Thames. Glancing at the water below us she said: 'You know the other great thing about granite? It's why dry land exists.' Around 4 billion years ago it was granite, because of its buoyancy, that rose from the Precambrian waters to form the continental (as opposed to oceanic) crust – the land that we live on.

During the nineteenth century rising wealth and the devel-opment of the railways made it easier to transport granite and other materials to the capital from around Britain: slate from Wales, sandstone from the Pennines, limestone from the Cotswolds. Nowadays, building stones tend to come from much further away. On our walk we found stones from Corn-wall, Dorset, Aberdeenshire and the Scottish Highlands but also from Italy, Greece, Norway, Sweden, China, South Africa and Australia. It's true that certain special stones have been brought in from overseas ever since the Romans, who trans-ported marble to the city: the green, purple and ochre marbles in the Brompton Oratory and the Art Nouveau interior of the Blackfriars Pub are a fine example of this tradition. But today it's also the everyday materials, such as granite for kerbstones, which come from so far away.

As the geologist and journalist Ted Nield writes in his book *Underlands*, the low price of oil and of foreign labour (with all that this implies about exploitation and poor working conditions) means that it is now much cheaper for a British firm to source granite from China or India than Aberdeenshire, and never mind about the additional fossil fuels consumed as part of that journey.[4] To complicate matters further, many of Britain's rocks now lie beneath areas designated as national parks and cannot be easily quarried. The situation turns into a paradox: the beauty of the landscape is preserved in the short term at the expense of the existence of the landscape, in its current form, in the future.

My walk with Ruth Siddall ended on the steps of St Paul's Cathedral. The nearby gardens were bright with daffodils and above us was that famous white façade, made of Portland Stone. The main landing on the steps of the cathedral is white Carrara marble with inner panels of red and grey Swedish travertine. If you crouch down to study the travertine, you can make out ghostly white markings, a segmented cone shape. This is an orthocone from Ordovician times, a marine creature related to the squid.

As we studied the fossil, I thought about how strange it was, and how unlikely, that this particular 440-million-year-old fossil had not only survived a journey through deep time, but had ended up one of just a handful of pieces of rock that would be selected by someone in seventeenth-century Sweden to transport to London to decorate the steps of a new cathedral, that it would later be part of Nelson's funeral in 1806 and

Churchill's in 1965, of Charles and Diana's wedding in 1981 and of the Occupy protests in 2012.

Tourists milled around in the afternoon sun and posed for photos on the steps. I was reminded again of Thomas Hardy's Henry Knight gazing at his trilobite: 'Time closed up like a fan before him. He saw himself at one extremity of the years, face to face with the beginning and all the intermediate centuries simultaneously.'[5]

*

In Italy the urban geology was different. 'A blindfolded geologist entering a brand new town, unknown to him, will have information on the local geology just by [looking at] the materials used in the buildings.'[6] So said the Italian mineralogist Francesco Rodolico, author of *The Stones of the Cities of Italy* (1953). I came across the quotation in a paper in the *Journal of the Virtual Explorer*, and was struck by the contrast between a city such as London and Rodolico's conception of a town, built of local materials. Certainly London has its traditional building stones but a geologist set down on the South Bank would struggle to explain the local geology using the Festival Hall. One of the paper's authors was Vincenzo Morra, and soon after my walk with Siddall, I travelled back to Naples to talk to him about the urban geology of the city.

In the old town, the historic centre of the city, dim and narrow streets run between tall buildings, laundry hangs from windows, scooters buzz everywhere and passageways open suddenly onto dusty sunlit squares. Churches with plain façades reveal interiors of rich marble and gleaming gold and silver. The university, founded in 1224 and counting Thomas Aquinas among its alumni, occupies a series of imposing buildings

in the historic centre of the city. It also contains the beautiful Royal Museum of Mineralogy (founded in 1801), with its
rows of glass-fronted wooden cabinets filled with 25,000 specimens from around the world. While we talked some of Morra's
colleagues came into his office and pulled out the espresso
machine, wedged among the piles of books. Someone mentioned the Rodolico quotation, and all the geologists started
talking about colours. A city built from local materials will
often have a specific colour scheme. Rome is white and red –
white from the travertine, a form of limestone, and red from
the bricks. Florence is white (marble), grey (Pietra Serena, a
sandstone) and green (serpentinite). Naples is dusty grey and
sandy yellow: Vesuvian lava, the rock known as Piperno and
Neapolitan Yellow Tuff. Morra stubbed out his cigarette in a
lumpy grey ashtray. 'Made of lava,' he said, 'from Mount Etna.'

Where the traditional building stones of London were
largely formed under water (Portland Stone, York Stone), the
stones of Naples come from fire. They are igneous rocks, volcanic in origin. From the window of the Museum of Mineralogy
you can see the towering blue peak of Vesuvius, which last
erupted in 1944. Morra and his colleague Alessio Langella both
live to the west, near Campi Flegrei. Langella's house is situated in the red zone – the area highlighted as being especially
at risk from volcanic activity – a fact that he and Morra seemed
to find particularly amusing.

When Vesuvius erupted in AD 79, it spewed out fiery, sticky
lava and clouds of burning volcanic ash. The town of Pompeii
was buried in a hot ash deposit up to 6 m thick, killing most
of the inhabitants. Walking with Morra and Langella through
the centre of Naples, I kept thinking about how very domesti-
cated all this fire and brimstone was. On a day trip to Pompeii
you can see those famous casts of ash-buried bodies, but in the

Mineralogy Museum are a series of jaunty souvenir medallions moulded from the lava of later eruptions, and on the streets of Naples the residents go about their daily business walking over dark grey pitted slabs of Vesuvian lava.

In a quiet courtyard, where busts of illustrious men are set among palm trees, we stopped to examine a weathered grey column of Piperno. 'This,' said Morra, 'is the product of the largest eruption in the Mediterranean for 200,000 years.' The Campanian Ignimbrite super-eruption, the volcanic event responsible for the formation of the Campi Flegrei caldera, created its own rock, the Piperno, which is found nowhere else in the world. Formed from compressed light-grey volcanic ash with fragments of black flattened scoriae (basaltic lava ejected from a volcano), also known as *fiammae*, it's a hard, heavy rock used occasionally for facing buildings – it gives Naples's Gesù Nuovo church its forbidding, fortress-like air – and more typically for portals and ornamentation. Sometimes, depending on the cut of the stone, the *fiammae* look like little black flames flickering over the surfaces of the buildings.

Among the streets to the north of the university, we passed Roman walls with diamond patterns picked out in pink bricks and yellow stones, and excavated blocks of sandy-coloured Greek masonry. For thousands of years people have been building here, often using the same rock: Neapolitan Yellow Tuff. Compared to the 2-billion-year-old South African gabbro on London's South Bank, the tuff is a young rock, a mere 15,000 years old. Like the Piperno, it is unique to this area and composed of compressed volcanic ash, but it is the product of a later, smaller eruption. It's also much softer, lighter and more easily cut – a good building stone, if protected from weathering – typically with a covering of yellow plaster that echoes the sandy colour of the stone itself. Where the plaster has fallen away, the tuff has a spongy appearance, filled with tiny holes that look like air bubbles. Near the Piazza San Giovanni Maggiore, I ran a finger lightly across an exposed block and it crumbled like the wall of a sandcastle. Alongside the grand churches and palaces and busy piazzas, you notice broken masonry, graffitied walls, small chapels closed for some indeterminably stalled restoration project. 'The most important problem for conservation in Napoli,' Langella said, 'is finding the money.'

Producers of wine – such as Chris White of Denbies in the North Downs – talk about the concept of *terroir*, an assumption that the soil, topography and climate of the land impart certain qualities to the grapes grown there. Siddall told me that she believed in a *terroir* for stones, and the idea seems to fit the architecture of Naples. While London builders had to send away for stones, the Neapolitans simply dug downwards, taking the tuff from underneath their city and piling it back up above ground in the form of houses, shops and public buildings. They left behind an immense network of passageways and around 2,000 caverns, a second city deep within the tuff.

Morra took me to meet a former student of his, Gianluca Minin, who is developing a partially abandoned complex of passages called the Galleria Borbonica. Formerly a system of Renaissance water tanks, a post-war rubbish dump and a 1970s police car pound, Minin has opened the space to the public as a sort of fantastical museum, art work, concert hall and adventure playground. As we walked through the tunnels, abandoned Alfa Romeos and crumpled Vespas loomed out of the darkness, all coated in a layer of fine powder, like something out of J. G. Ballard – a vision of the death of modernity. 'In 2005 the government sent me to survey the caves here and I fell in love, as, you know, it is possible to fall in love with a woman,' Minin said, gesturing expansively towards a space where Neapolitans had sheltered from Allied bombing raids during the Second World War. 'I wanted to save it all, to preserve the history.'

In their paper in the *Journal of the Virtual Explorer*, Morra and Langella wrote with concern about the declining use of traditional local building materials such as the Neapolitan Yellow Tuff.[7] It's a familiar story. If the landscape near Naples was to be preserved, the quarrying near Naples had to cease. Morra and Langella favour a limited return to quarrying in the region, using modern, less invasive techniques, in order to provide materials for restoration and the construction of significant architectural structures. In London, Siddall had worried about building materials being shipped from China instead of, say, Aberdeen. Near the Piazza Luigi Miraglia, Langella pointed out a new-looking grey column. 'Lava, but it's not local,' he said disapprovingly. 'It's from Mount Etna.'

Before leaving Naples, I walked back to the Piazza San Giovanni, passing a church where plants with small, pinkish flowers were growing out of the wall high up near the bell tower. In the piazza, students with heavy boots and metal piercings sat on the paving stones talking and smoking. The lava had a glassy aspect, polished by generations of pedestrians. How many generations? How many years? And how many more years since the limestone of Waterloo Station had formed, or the gabbro of the Nelson Mandela plinth.

Sitting outside a café at the edge of the piazza, I looked over my notes. 'It was Rodolico who first talked about the relationship between geology and building stones,' one of Morra's colleagues had said. 'Here in Naples now, we are building very strong relationships with architects and engineers to try to solve the problems of decay, of the conservation of buildings.'

Slowly, I became aware of a series of people in white overalls, entering the piazza in twos and threes. They gathered in the centre and unfurled hand-painted banners. Someone brought out a video camera. I walked over and asked what was happening. A woman explained that some years ago the European Union had awarded a large sum to restore Naples's historic centre: 75 million euros, I later read.[8] The crowd, mostly out-of-work art and architectural restorers, were excited but anxious. There were rumours of delays, bureaucratic incompetence, Mafia connections, perhaps malpractice. They wanted now to draw public attention to the project to ensure that it went ahead.

'There is so much work to do,' the woman said, bending down to pick up one end of her banner and waving it above our heads.

14

IN SEARCH OF THE ANTHROPOCENE

In the city in high summer the sun presses down on your skin, heat hangs thickly in the spaces between the buildings and the air is full of dust. On the news sites there are pictures of office workers laid out on the browning grass of Green Park.

Online I read that across the 4.6 billion years of deep time geologists have found evidence of many distinct periods of global warming. Fifty-five million years ago, for example, the Earth endured a climate event known as the Palaeocene-Eocene Thermal Maximum, or PETM. Geologists studying the rocks can tell that massive amounts of carbon were injected into the atmosphere, and that there was a sharp increase in global temperature. 'The PETM event wiped out a lot of mammals,' Ruth Siddall told me. 'Mammalian evolution had to start again after that, so we know that it's not good for us mammals, a heating climate.' On the ten o'clock news a climate change protester carried a banner: 'Mother Earth Needs You.' I thought back to something else Siddall had said. We may need the Earth, but 4.6 billion years of history shows that the Earth doesn't need *Homo sapiens*: 'The Earth will recalibrate and something else will appear, but it won't be us.' I also thought about what a geologist had told the writer John McPhee back in 1981. How had living in deep time affected him? McPhee asked. 'You care less about civilisation,' the geologist said. 'Half of me gets upset with civilisation. The other half does not get upset. I shrug and think, "So let the cockroaches take over."' [1]

Until recently our current interglacial was scheduled to end in around 50,000 years. Vast emissions of greenhouse gases seem to have put paid to that.[2] 'When I was a kid, the talk was about how the next glaciation was coming,' John Marshall had said to me. 'If you look at the novels of the '70s, they've got New York frozen solid. Now it's always flooded.'

Such a decisive warming of global temperatures is a piece of the evidence put forward by the scientists who argue that we have now moved from the Meghalayan into the Anthropocene – the proposed new geological unit that says that humans are altering the planet, including long-term global geological processes, at an increasing rate, and that we have already changed the Earth system sufficient to produce a distinct stratigraphic signature in sediments and ice. At the time of writing, the International Commission on Stratigraphy has yet to come to a decision on the official creation of the Anthropocene as a stratigraphic unit, but, as the British Geological Survey scientist Colin Waters writes, 'Not only would this represent the first instance of a new epoch having been witnessed first-hand by advanced human societies, it would be one stemming from the consequences of their own doing.'[3]

In Copenhagen, when I asked Jørgen Peder Steffensen about the Anthropocene, he said:

> You can assume scientifically that anything pre- the Anthropocene can be explained as a consequence of a link of events leading on to that moment. You cannot say that in the Anthropocene. Of course the physics is the same,

but the mechanism behind the change is actually a conscious mechanism making decisions, and that is a break. The Anthropocene is the only period where you cannot just explain things using unbiased scientific causality: it is the only period that is influenced by the human mind.

In addition to global warming, examples of human influence might include: modified global cycles of carbon and nitrogen; rates of extinction well above background levels; enhanced erosion; and a spike in artificial radionuclides caused by thermonuclear weapons tests in the mid-twentieth century.[4] Viewed through the lens of the Anthropocene, humans become a geological force comparable to immense volcanic eruptions such as the Campanian Ignimbrite, or the variations in the Earth's orbit that drive glacial cycles.

But what does it mean to become a geological force? Like the immensity of deep time, it's something you can know intellectually but not emotionally or bodily. Dipesh Chakrabarty, a historian from the University of Chicago who has written about the Anthropocene, told me that

One of the problems evolution leaves us is we have a big brain that helps us understand very large, abstract events, but also leaves us with the level of human experience that phenomenologists study, which is your internal time consciousness, your internal awareness of your body. This is what creates your social sense of who you are, your sense of ego. It's extremely important – it's what gives meaning to our lives – and it is predicated on the span of individual human lives and the inevitability of death. Our being anchored in this short span, phenomenologically, is something that you can't avoid, even when you think about very large problems that can continue for hundred of generations. That's why, when it comes to making policies and taking political decisions, you always do it in the short term. We think immediately, what does it mean for me and for my children?

If our sense of self could be enlarged, if we could feel – viscerally feel – things on a deep, not shallow, time-scale, then perhaps it would not be so extremely difficult to break out of our current patterns of thinking, which are always short-termist, which mean that many of us drive a car to the shops, or take a plane to travel abroad, when we know logically that such actions are indefensible.

In the summer of 2015, working on a piece about the Anthropocene, I had listened to Jan Zalasiewicz, chair of the stratigraphers' Anthropocene Working Group (AWG), speaking at Tate Modern alongside architects, historians, philosophers and artists as part of the Anthropocene Project conference. Leaving the stage, he quickly found himself at the centre of a crowd of artists and curators who wanted to talk about the Anthropocene. Men in black-framed glasses and women with severe fringes discussed tectonic forces and the infrastructural unconscious of contemporary urbanism. Next to me, someone said they preferred the term 'Capitalocene', acknowledging the hugely unequal contributions towards CO_2 levels from 'developed' and 'undeveloped' nations. To call it the Anthropocene, they argued, was unfairly to implicate the entirety of humanity. The guilty party was the capitalist system or the world global economic order.

Chakrabarty, also at the conference, suggested that 'culturally, what the term does is, it invites you to place human beings in the context of deep time.' In the Anthropocene, histories usually kept separate are brought together. The burning of fossil fuels during the Industrial Revolution is shown to be simultaneously a part of human history and the history of the Earth. For the first time in human history we are consciously connecting events that happen on vast geological scales – such as changes to the climate system – with what we do in our everyday lives, such as burning fossil fuels. 'We are undergoing a profound shift in what I'd call the human condition,' he said.

A shift in the human condition might mean trying to re-imagine ourselves as a geological force – something capable of altering the usual cycle of the ice ages or creating a sixth mass extinction event. It might mean entering some sort of new, mythic space where our domestic activities – switching on the washing machine, driving to the supermarket, discarding a plastic food container – take on a cosmological significance. Simply by living our everyday lives we build a new Earth around us.

*

'Concrete is a new kind of rock. There has been nothing quite like concrete in Earth's 4½-billion-year history, and by now we've made about 500 billion tonnes of it, which is enough for one kilo for every square metre of the Earth's surface, land and sea.'

Zalasiewicz and I were standing in the middle of the University of Leicester's campus, contemplating the Anthropocene. Concrete, he explained, can be considered a characteristic deposit of the proposed Anthropocene Epoch, just as the Carboniferous is a time inextricably linked to coal or the Cretaceous to chalk. '[Paul] Crutzen and [Will] Steffen and others looked at the Anthropocene in terms of chronology, history, events which they can observe and so on, but as geologists we also have to look at the Anthropocene in terms of the record it will leave behind as strata,' he said.

Zalasiewicz has a gentle, infectious enthusiasm for his subject. In the future, he has written, the ethical implications of the Anthropocene 'may stimulate transformational thinking, to enable us to better integrate into the Earth System'. This new geological unit could one day promote 'hope rather than despair'.[5] Besides chairing the AWG, he studies Ordovician graptolites and in his spare time writes acclaimed popular science books and articles on subjects ranging from the geology of First World War battlefields to Mary Anning and the Comte de Buffon. 'I think about the eighteenth- and nineteenth-century naturalists and how they came at everything fresh. We come at it with clutter, with all this bloody education,' he said. For Zalasiewicz the Anthropocene is, among other things, an opportunity to see anew what has always been in front of us: 'We can look at buildings, for example, and say that they are also part of the rock cycle. They are made of rocks and will go back to being rocks. They will leave a distinctive record.'

Later, waiting at King's Cross for a Victoria line train, I peered down the tunnel of pre-cast concrete rings to watch for the glowing lights of the tube trains. Seen through the prism of the Anthropocene, everything changes. It's our world, but not quite. Our world as seen through the eyes of someone not from our world. A concrete tube station becomes a new rock, a tube tunnel an example of bioturbation – the disturbance of strata by living organisms. One day the tunnel may become a trace fossil, a mark left behind in the rock — like a dinosaur's footprint — showing us that something living once passed that way.

The deepest animal burrowers under normal circumstances are thought to be wolves and foxes – which dig down to 4 m, though Nile crocodiles may retreat up to 12 m below ground during dormant periods, where they shelter from the worst of the sub-Saharan African heat. Plant roots in the Kalahari Desert have been recorded at depths of 68 m. But the extensive, large-scale disruption of rock to depths of more than 5 km by a single biological species represents a major geological innovation with no analogue in Earth's 4.6-billion-year history.[6] Below our daylight world is another dimmer, greyer world, made up of sewage, electricity and gas systems, underground metro lines, repositories for nuclear waste, mines, wells and boreholes. We are the only species to stray so far beyond our own domain, our tunnels and pipelines disturbing the ancient layers of strata that record the story of our planet stretching back into the vastness of deep time.

And this crepuscular second world will probably long outlive the daylight one where we spend most of our time. Zalasiewicz and colleagues write that these buried structures, 'being beyond the immediate reach of erosion, have a much better chance of short- to medium-term preservation than do surface structures made by humans'. If an underground structure is in a place being raised up by tectonic forces, it will eventually break the surface and be subject to erosion – though burial of a few kilometres, for instance, means that this probably won't happen for millions or tens of millions of years. 'Structures on stable or descending crust may of course remain preserved below the crust for very much longer, or even permanently.'[7]

Human bioturbation can also be seen as a part of what Zalasiewicz and colleagues have called the 'technosphere'. The physical technosphere is defined as 'the summed material output of the contemporary human enterprise'.[8] It is made up in part from things with the potential to become 'technofossils' – the fossilised trace of a manufactured product. A palaeolithic flint axe might become a technofossil, as might a tube train, road, power station, ballpoint pen or toothbrush.

Zalaseiwicz was excited: 'One thing we want to pursue is, what is the true diversity of techno-fossils? Has anybody ever counted how many different types of toothbrush have been made?' So far, the answer to that question is no, but Zalasiewicz has, for example, used Google to calculate that around 130 million individual book titles have been recorded since publishing began, 'with now over a million new titles each year in the USA alone'. He has written that 'each title can be regarded as a biologically produced, morphological entity with its own specific pattern of printed words, dimensions and texture.'[9] This is much the same way that we identify an ammonite, say, from its size and the patterns on its shell, though as with biological species, much information will necessarily be lost. Zalasiewicz expects the books to survive, if at all, as 'rectangular carbonised masses classifiable by size and relative dimensions and subtle variations in surface texture; fragmentary details of the print information will only be rarely preserved, as are fragmentary details of DNA structure in some exceptionally preserved ancient fossils today.'

*

So if we are in the Anthropocene, when did it begin? According to the AWG, any proposed date needs to show the first appearance of a clear synchronous signal of the transformative influence of humans on key physical, chemical and biological processes at the planetary scale.

Paul Crutzen argued for a start date of AD 1784 to coincide

with the invention of James Watt's steam engine and the beginning of significant increases in CO_2 emissions related to the Industrial Revolution.[10] Other suggestions include a date linked to the expansion of agriculture and livestock cultivation during the Neolithic, or one that coincides with the growth of mining around 1,400 BC. (A difficulty with using these events is that they occurred at widely different times in different places around the world.) A paper by Mark Maslin and Simon Lewis proposed a start date of 1610, when there was a noticeable drop in global levels of CO_2.[11] The authors linked this to the deaths of some 50 million indigenous people in the Americas, triggered by the arrival of Europeans. Abandoned agricultural fields were overrun by new forest growth, and this led to the drop in CO_2.

In May 2019 the AWG homed in on a mid-twentieth-century start date. This would coincide with the so-called 'Great Acceleration' in human population, resource consumption, global trade and technological evolution. After this time period, a scientist sampling, say, a lake sediment core might expect to find radically different results from those from the Meghalayan. These new results might include 'unprecedented combinations of plastics, fly ash [a by-product of burning fossil coals], radionuclides, metals, pesticides, reactive nitrogen and consequences of increasing greenhouse gas concentrations'.[12] Elsewhere, changes to the fossil record would also be evident – this organism stopped existing; this one suddenly appeared on the other side of the world (i.e., it was moved by humans). AWG members voted by 29 to 4 to formally define the Anthropocene as a stratigraphic unit beginning in the middle of the twentieth century. They are currently searching for a suitable GSSP or 'golden spike' to mark this mid-twentieth-century date. Their findings will be presented to the Subcommission on Quaternary Stratigraphy and then, if they are approved, the International Commission on Stratigraphy.

Zalasiewicz is cautious when asked about his chances with the ICS: 'It's a very conservative body, but the right questions are now being asked ... In my personal opinion there's no question that we've entered the Anthropocene.'

*

Not everyone within the geological community is so convinced. 'The Anthropocene? It's not old enough for me to really care about,' the BGS sedimentologist Romaine Graham had told me. 'It's geography, not geology.'

'If it makes everyone realise that we are impacting on the environment big time, and do something about it, then that's no bad thing,' Andrew Farrant said. 'But I do think its one of those things where the press gets hold of something and it gets built up into a far bigger thing than it needs to be. If someone came up with a new group stage in the Jurassic between the Toarcian and the whatever, it wouldn't cause a ripple except with us lot.'

Meanwhile Stan Finney, formerly chair of the ICS and no fan of the Anthropocene, has argued that other organisms have had far greater longer-term impacts on the planet that have not been recognised with a formal time unit. Consider the evolution of vascular land plants and their spread across the continents from the Devonian to early in the Permian. This dramatically altered CO_2 and O_2 concentrations in the atmosphere and oceans to a 'far greater [level] than humans are projected to do', Finney writes in a challenge to the AWG.[13] 'Might the desire to establish the "Anthropocene" as a formal unit be Anthropocentric?'[14]

At university in Sheffield, Zalasiewicz was a few years behind Philip Gibbard, though they didn't know each other then, meeting later and becoming friends while working on the East Anglian ice sheet. As chair of the Subcommission on Quaternary Stratigraphy, Gibbard asked Zalasiewicz to set up the Working Group, but still he worries about 'this Anthropocene business'. He told me: 'I never thought I'd be cast in the role of the man with his foot on the brake, but I do think we need to say "Slow down" ... Certainly there have been huge amounts of anthropogenic change – and when the Green lobby say that we're a bit of a plague on this planet I tend to agree with them – but does it warrant a change to the [ICS] chart?' Gibbard worries that the geologists have lost control of the narrative, and that complex science is being misrepresented: 'In many ways this is exactly like the dinosaur business all over again – the whole thing grossly oversimplified by rocks falling from the sky et cetera et cetera.'

Crutzen, not himself a geologist, saw the Anthropocene concept as fundamentally interdisciplinary, and while the AWG is made up predominantly of geologists, it also includes environmental scientists, archaeologists, a philosopher, a lawyer and a historian. This makes Gibbard – also a member of the group – anxious. 'However other people want to use the term "Anthropocene" in their own disciplines is fine – if it forms a function for them that's fine – but whether we geologists should define it and add it to the chart depends on its usefulness for geology.' Certainly the Anthropocene question flips the usual practice of stratigraphy upside down. Traditionally a geologist looks at some rocks and works out something about the events of Earth history. With the Anthropocene the events are already known and documented through human observation, and the geologist is asked to search for evidence in the rocks.

Gibbard has written, as Maslin and Lewis argued with reference to Meghalayan debate, that 'it is the "anthropogenic" signature that is the hallmark of the Holocene, setting it apart from previous interglacials.'[15] He, however, draws different conclusions from the geographers. 'Now from my perspective as a geologist I would say you can't play the same card twice, and so that means that what's been going on to establish the Holocene – the presence and activity of *Homo sapiens* – can't then be used again to define the Anthropocene,' he told me.

In response, Zalasiewicz argues that this equation is not explicit within the formal definition of the Holocene, that humans were also a component of the Earth system in the late Pleistocene – i.e., before the Holocene – and that in any case the critical point 'is not the fact of any human impact but fundamental differences between the Holocene and the Anthropocene as regards the magnitude, rate and global synchronicity of change recorded in their respective stratigraphic signatures'.[16] The Anthropocene would still be geologically distinct were the drivers of the transformative processes cats, say, instead of humans.

Eric Wolff, also from the University of Cambridge, has suggested that 'it may be more sensible to describe ourselves as being firmly in the transition into the Anthropocene, but to leave future generations to define, with the benefit of hindsight, when the epoch actually starts or started.'[17] Gibbard agrees. 'In many ways, we are just too close to the events to be able to talk usefully about them from a geological perspective. Even the Holocene is very short. It begins 11,700 years ago, and that's nothing, just a tiny period, a blink of the eye. Geologically, even forty-five years ago is still "now".'

*

Between 2000, when the term was first conceptualised by Crutzen as a new and distinct unit of Earth history, and the end of 2017 the term 'Anthropocene' was used in more than 1,300 scientific papers, which collectively have been cited more than 12,000 times.[18] It has given rise to at least four scientific journals and is in the title of more than 100 books. When I started reading about the Anthropocene, back in 2014, few people I spoke to had come across the term; now it typically raises at least a nod of recognition. The concept has spread with extraordinary speed across the sciences to the humanities and the arts, becoming the subject of photography, poetry, Pinterest fashion boards, opera, Death Metal albums and a song by Nick Cave.

So what is it that makes the Anthropocene narrative so compelling? Certainly it offers a gripping vision of the magnitude of humankind's impact on the planet at a time when green issues are becoming ever more visible and politically resonant. And it conveniently gathers and labels familiar narratives of planetary disaster and the looming threat of extinction (though it should be noted that this description presents the Anthropocene as a wholly negative condition, which is not the premise of the scientists' work). But is there another side to our fascination with the Anthropocene?

A literal interpretation of the Bible places humans at the centre of creation. In the sixteenth century Copernicus dislodged us spatially, showing that it was the Earth that circled the sun, not vice versa. During the nineteenth century Charles Darwin – making use of Charles Lyell's work on, among other things, the age of the Earth – showed that we were not at the pinnacle of the tree of creation, merely a branch alongside other branches. Later work by Arthur Holmes and others began to make apparent just how incredibly slight was *Homo sapiens*'s share of the vastness of deep time.

Considered one way, then, the Anthropocene concept puts humans back at the centre of the world – the place where, for hundreds of years, we unquestionably thought we were. And at some level we can't help finding that attractive – even if the price for that return is environmental disaster.

15

'THIS PLACE IS NOT
A PLACE OF HONOR'

This place is not a place of honor.
No highly esteemed deed is commemorated here.
Nothing valued is here.
What is here was dangerous and repulsive to us.
This message is a warning about danger.
Kathleen M. Trauth, Stephen C. Hora, Robert V. Guzowski,
Sandia National Laboratories report, 'Expert Judgement
on Markers to Deter Inadvertent Human Intrusion into the
Waste Isolation Pilot Plant'

Mid-January in the small town of Rauma, on the west coast of
Finland. It was six in the morning and dark outside, but when
I pressed my face to the window I could see snow falling and
banking up in the square outside my hotel. The evening before,
Pasi Tuohimaa, head of communications for nuclear power
company Teollisuuden Voima Oyj, had shown me around.
Tuohimaa had been brought up in the town and then moved to
Helsinki to become a journalist and TV presenter before switch-
ing to PR, which paid better but meant living in Rauma during

the week, only returning to Helsinki and his family at weekends. I got the impression that he didn't have many friends left in Rauma and was lonely. As we walked the empty streets he pointed out his old school, the rink where he played ice hockey, the basement where his band used to rehearse. When I told him how much I enjoyed the films of Finnish director Aki Kaurismäki, he made a face. All foreign journalists wanted to talk about Kaurismäki. Kaurismäki was not a good representation of Finland.

Rauma has a population of around 34,000. It is famous for lace-making and colourful wooden houses. It's also around four miles from the island of Olkiluoto, where three of Finland's five nuclear power plants are situated, along with the thing I'd come to the country to see: a network of caverns and passages sunk 450 m into the Finnish bedrock and known as Onkalo.

On the morning of my visit the hotel breakfast buffet was full of construction workers and engineers bound for Olkiluoto. Men in fluorescent trousers helped themselves to plates of scrambled egg and rye bread. When I asked the hotel receptionist whether Rauma's proximity to Onkalo bothered her, she shrugged and shook her head. 'It's just normal,' she said.

But Onkalo – the names means 'cavity' or 'cave' in Finnish – isn't normal. It has been designed not for human but for deep time. The men and women building it say that it will last for more than 100,000 years. So far, no man-made structure has existed for that amount of time. (The oldest Egyptian pyramids are only around 4,800 years old, Stonehenge perhaps 4,500.) In deep time, 100,000 years is long enough for the world to change utterly. A hundred thousand years before now, Europe was in the grip of an ice age, modern humans had not yet arrived on the continent and mammoths and woolly rhinoceroses roamed the landscape. Where will we be 100,000 years hence?

Onkalo is the world's first geological disposal facility for high-level nuclear waste – material so potent it must be isolated from humans for at least 100,000 years – some say a million – until the levels of radiation, which decrease over time, are no longer deemed hazardous. If everything goes according to plan, it will be one of the things we leave behind us. Something man-made that will go on persisting into the deep future. Onkalo may very well be our legacy.

*

Onkalo is being built on a flat stretch of land covered by pine trees and bordered on three sides by the Baltic Sea. On the fourth side a small channel separates the island from the main-land. The taxi driver followed a narrow road between tall pine trees that suddenly opened out to reveal a glass-fronted visitor centre. It was still dark, and a strong breeze was blowing snow off the roof of the centre and off the branches of the pine trees. In the distance, across the black channel of water, the lights of the nuclear power plants winked on and off.

The brightly lit visitor centre was open but deserted – worried about being late, I had arrived too early for my inter-view with Tiina Jalonen, Senior Vice President of Development for Posiva, the company founded by TVO and Fortum Power & Heat Oy to manage their nuclear waste disposal. To pass the time I wandered around the display in the foyer, where a sunken-eyed animatronic Einstein model tried to demystify the business of nuclear energy.

We are exposed to radiation every day. Public Health England estimates that in a typical year a person in the UK might receive an average dose of 2.7 millisieverts (mSv) from natural and artificial radiation sources. A transatlantic flight, for example, exposes you to 0.08 mSv, a dental X-ray to 0.005 mSv, eating 100 g of Brazil nuts to 0.01 mSv.[1] On average, Americans are exposed to more radiation than Britons, in part because they often have more medical scans.

In a nuclear power plant, nuclear fission – the process of splitting the nucleus of an atom of radioactive uranium or plutonium – is used to heat water, produce steam and drive turbines. At the Olkiluoto plants the spent fuel that results from this process consists of dark brown ceramic pellets of uranium oxide sealed into zirconium alloy tubes known as fuel rods. A single uranium fuel pellet is about the size of a small fingernail, and four of these pellets contain enough energy to provide a year's electricity for a family of four living in an electrically heated house.

The displays were more reticent about the dangers of radiation. The nuclear industry, you sense, is aware that it has an image problem. Stand in front of an unshielded source of radiation and you won't see or feel anything. However, some of that radiation will be passing into your body. Nuclear waste is dangerous because it emits ionising radiation in the form of alpha and beta particles and gamma rays. While alpha particles are too weak to penetrate the skin, beta particles can cause burns. If ingested, both can damage internal tissues and organs.

It's gamma rays, however, that have the greatest penetrating range, and therefore the potential to cause the most widespread damage to the DNA of your cells. This damage may lead to an increased risk of cancer later in life, and is largely responsible for the set of symptoms known as radiation sickness. Some experts estimate that a dose of over 1 sievert is enough to cause radiation sickness. Symptoms include nausea, vomiting, blisters and ulcers; these may begin within minutes of exposure or be delayed for days. Recovery is possible, but the higher the radiation dose, the less likely it is. Typically, death comes from infections and internal bleeding brought about by the destruction of the bone marrow.[2]

Spent fuel has been described by nuclear researchers as 'the Achilles heel of nuclear power'.[3] It's a problem for all of us – whether pro, anti or indifferent to nuclear – because even if all the world's nuclear plants were to cease operating tomorrow, we would still have more than 240,000 tonnes of the stuff to deal with. This figure will of course increase, especially as new power plants such as the UK's Hinkley Point C come into service, and as going nuclear is increasingly viewed as a way of cutting carbon emissions. The UK's Nuclear Decommissioning Authority estimates that by 2125 the UK alone will have produced around 1,500 m^3 of spent fuel.

At the moment all nuclear waste is stored above ground or very near the surface in interim wet or dry storage facilities. This has never been considered an acceptable long-term solution. In the past, ideas for dealing with the waste have included shooting it into space, burying it in deep ocean sediment and, incredibly, dropping it into a gap between tectonic plates. There is some interest in the use of deep boreholes, but according to Rebecca Tadesse, head of Radioactive Waste Management and Decommissioning at the Nuclear Energy Agency (NEA): 'Internationally it's been recognised that geological disposal is scientifically the optimal approach.' At the time of writing, SKB, the Swedish Nuclear Fuel and Waste Management Company, have completed the first part of the licensing process for their site, and Andra, the French National Radioactive Waste Management Agency, is putting together a planning application for their own repository, Cigeo. The UK, Canada, Germany, Switzerland and Japan are all searching for suitable sites, while the USA continues to debate a potential facility in Nevada's Yucca Mountain.

In the visitors centre Tiina Jalonen appeared wearing,

despite the snow, a pair of high black stilettos. 'I think that the generations that have produced the nuclear waste should take care of it,' she told me, sitting down. 'We shouldn't leave it to our kids.'

Above-ground storage facilities will require active monitoring for hundreds of thousands of years. As well as regular refurbishment they must be protected from, for example, earthquakes, fires, flooding, deliberate attacks by terrorists or an enemy nation. This not only places a financial burden on our descendants, who may no longer even use nuclear power, but also assumes that in the future there will always be people with the knowledge and will to monitor the waste. On a 100,000-year time-scale this cannot be guaranteed. 'We can't always control conditions above ground,' she said.

Not everyone agrees with this assessment. Peter Roche, who formerly worked on nuclear issues for Greenpeace and is now a policy adviser for the local government organisation Nuclear Free Local Authorities, voiced the concerns of some anti-nuclear campaigners when we spoke on the phone before I travelled to Finland: 'It's the uncertainty of it all that worries us.' Can you ever really know how a certain material will behave in a certain location in 100,000 years' time? 'If you end up putting it down a deep hole and it starts to leak, then you've left future generations with something they can't do anything about.'

Others, such as Gerry Thomas, Chair in Molecular Pathology at Imperial College London, are hesitant not because they think geological disposal facilities are unsafe but because they believe that in the future we may have other uses for the waste. 'Burying something that's a potential fuel that is not carbon – is that a really good idea? Repositories make sense if you just want to put it in a hole and walk away – and of course that makes us feel better about it – but scientifically there are big questions around whether that's actually a misuse of future energy.'

Her comments made me wonder if there isn't something a little Freudian about this desire to hide our waste by burying it out of sight underground. For Thomas, a woman who spends a lot of time telling journalists that radiation is not scary, it's also a PR own goal. 'I can see why they want to do it because everybody's scared of radiation, but actually they're helping to perpetuate the fear.' The nuclear industry has always struggled to throw off the taint of its secretive, shadowy beginnings within the military. For many of us, nuclear is still Hiroshima, Nagasaki and Chernobyl. It's Godzilla and the green, glowing Hulk. Growing up in the 1960s, Thomas said, she was made to fear the atomic bomb and this spooky substance you couldn't see or touch but that could kill you. 'We have built radiation into this awful spectre of death, which it isn't, and we can't deal with that emotionally as a society.'

*

Past the pine trees, past a huddle of yellow, grey and blue huts, you reach the centre of the island. Here, grey rocks have been blasted open and the road slopes downwards between high, craggy walls towards a metal shutter set into the rock. This is the entrance to Onkalo.

The only way for an outsider to gain entry is as part of an official tour, such as a press event. There were five of us – three writers and two photographers. Tuohimaa handed out torches, oxygen tanks and injunctions against photographing the entrance. Security reasons, he said. Also for security reasons we would be accompanied at all times by a silent, bearded and unsmiling security guard.

In a minibus the long descent down the access tunnel takes twenty minutes. Intermittent electric lights gave out a dim glow. Cables and tubes snaked along the ceiling. Incomprehensible symbols were scrawled in fluorescent paint on the walls. I tried to remember the plan that I'd studied online, the way the large access tunnel looped down and round like a lazy helter-skelter. It was a weekday, but during our decent we saw few signs of life – only once, two men in fluorescent overalls standing where a tunnel branched off into the darkness, glancing up to watch the lights of our vehicle going past.

When we finally got out of the minibus, everything was very quiet except for a buzzing noise that might have been the lights or the ventilation. We were now at the planned disposal level 420 m below ground, where the dense, impermeable rock is most stable and least fractured. I walked over to the wall of the tunnel. Immensely, impossibly old, this metamorphic rock is a type of migmatitic gneiss that formed some 1.9 billion years ago from even more ancient sediments and volcanic rocks subjected to extreme heat and pressure. It is grey, shot through with white bands, and up close, at the right angle, it appears to sparkle.

As I stood there, the rest of the group walked a little way off to examine a test hole. There was no one else around, only a large, unmanned excavating machine, an abyssal creature looming out of the darkness. I felt a sudden eerie slippage as though in this timeless underground darkness it might as easily be some point hundreds of thousands of years into the future, and what I was seeing the remains, not the construction, of Onkalo.

I thought about all the people working on the site. Engineers, construction workers, scientists, communication officers, administrators, payroll managers. The Anthropocene problem could be framed as the problem of getting people to attend to something that neither they nor their children, nor even their grandchildren, will be alive to see. The workers at Onkalo are engaged on a project that will go on after their deaths, like some secular version of the medieval builders who, over successive generations, produced the great cathedrals of Europe knowing that they would never live to see their work completed.

If all goes well, the process for final disposal will be as follows. Spent fuel rods are taken from the nuclear reactors and cooled for several decades in pools in an interim storage facility. During this time both the temperature and the radiation levels decrease. Bundles of spent fuel rods are then placed inside 1-m-wide cylindrical cast-iron canisters, the material selected for its strength and ability to resist mechanical stress.[4] Once the waste is below ground, the major threat to public health comes from water contamination. If radioactive material from the waste were to mix with flowing water, it would be able to move relatively swiftly through the bedrock and into the soil and large bodies of water such as lakes and rivers, finally entering the food chain via plants, fish and other animals.[5] To guard against this, the iron canister is placed inside a 5-cm-thick copper canister that protects it from the corrosive effects of groundwater. (Reserves of natural copper have shown that copper can last hundreds of thousands of years in the bedrock without corroding.) The canister is placed in a cylindrical hole padded with bentonite clay, which, in the presence of water, will expand and solidify to protect the canister from rock movements. You can see this 'clumping' effect for yourself: bentonite clay is commonly used as cat litter. The clay also stops liquid from reaching the canister and, if there were a canister leak, would stop radioactive substances from reaching the rock. The rock itself, because it is dense and impermeable, acts as a further barrier.

In the planning diagrams, the completed disposal tunnels look like a series of grills, with each tunnel forming a bar of a grill. Disposal of the spent fuel will begin in the 2020s and continue for around 100 years. By the time the final disposal is completed in the 2120s, some 2,800 canisters will have been deposited in around 60–70 km of tunnels. They will contain around 9,000 tonnes of spent fuel – roughly enough to fill two Olympic swimming pools. No one working on the project today will be alive to see this.

The day after my trip to Onkalo I went to visit Jussi Heinonen, director of the nuclear waste and material regulation department at STUK, the Finnish Radiation and Nuclear Safety Authority. When I mentioned Roche's concerns about the limits of our technical knowledge, Heinonen nodded and sighed. 'It is true that the spent fuel is harmful for a very long time, but what we are maybe not explaining so well is that the harmfulness changes.' The level of radioactivity exponentially decreases, with the first 1,000 years being the most dangerous period. 'We are asked, how can you know how material behaves for several tens of thousands of years or even longer? And of course in that you need to be humble and say, well, we can't say we know everything. But the first thousand years are the most important, and we can be much more confident about what happens during that shorter time period.'

Like Jalonen, he believes that the time has come to jettison interim storage facilities: 'With interim storage you are relying on the fact that in several hundred years' time we will still have some form of Finnish society taking care of the waste. That's just not really something we can trust in.'

*

Some time after my visit to Onkalo I travelled to Paris to visit the headquarters of the NEA and Andra, the French nuclear agency, and to the Meuse/Haute-Marne region of rural north-eastern France to the proposed site of the French repository.

The safety of the waste relies on its isolation from humans and other animals. The strength of a deep geological repository is that it is designed as a passive system, meaning that once Onkalo or Cigeo is sealed, no further maintenance or monitoring is required. Much more difficult to plan for is the risk of human intrusion, whether inadvertent or deliberate. Imagine, one argument goes, that hundreds of thousands of years into the future all knowledge of the repository has been lost. Imagine a society that has forgotten the science of radio-activity. Imagine that one day someone discovers a strange slab of concrete buried in the forest floor ... How do we make sure the people of the future don't dig the stuff up and die some grisly, sci-fi movie death?

'Or what if somebody just digs into the ground near the repository?' asks Andra's Jean-Noël Dumont. Such a hole might allow flowing water to come into contact with the waste. According to Dumont, Andra's safety studies have taken this scenario into account, and any environmental impact is 'acceptable', but 'it would be better if there was no impact'.

How, then, to ensure that our descendants – assuming that there are still people living 10,000 years, 100,000 years, in the future – know about the dangers of the material buried at Onkalo, Cigeo and the other proposed repositories?

'All countries are considering which is the best way to maintain information about the repository for future generations,' Tadesse said. The French memory programme, however, is the most visible and most advanced. Dumont is in charge of this division. He works on the premise that information about the repository should be maintained for at least one million years. 'To my knowledge none of our counterparts has dedicated a really specific programme to the question of memory,' he told me. As he sees it, such a programme is necessary for three reasons: first, to avoid the risk of human intrusion by informing future generations about the existence and content of Cigeo; second, to give future generations as much information as possible to allow them to make their own decisions about the waste (for example, if Thomas is right that much of the waste destined for repositories may one day provide an important fuel source); and third, cultural heritage – a properly documented geological repository would provide a wealth of information for a future archaeologist. 'I have no knowledge of other places or systems where you have at the same time objects from the past and a very large, concrete description of how these products were manufactured, where they come from, how we considered them and so on,' Dumont said.

The history of memory programmes – memories of memory programmes – goes back to America in the 1980s, when the US Department of Energy created the Human Interference Task Force (HITF) to investigate the problem of waste repositories and human intrusion.[6] What was the best way to prevent people entering a repository and either coming into direct contact with the waste or damaging the repository, leading to environmental radiological contamination? Over the next fifteen years a wide variety of experts would be involved in this question, including materials scientists, anthropologists, architects, archaeologists, philosophers and semioticians – social scientists who study signs, symbols and their use or interpretation.

Suggestions generated in response to the HITF sometimes sound more like sci-fi than science. Stanisław Lem (in fact, a sci-fi author) suggested growing plants with warning messages about the repository encoded in their DNA. The biologist Françoise Bastide and the semiotician Paolo Fabbri developed what they called the 'ray cat solution' – cats genetically altered to glow when in the presence of radiation.[7]

Quite apart from the technological challenges and ethical issues these solutions present, both suggestions have one major drawback: to be successful they rely on additional external, uncontrollable factors. For the plants, you need to assume that our descendants will have the technology to read DNA codes and that they will bother to sample the DNA of some very specific plants in one very specific location. In the case of the cats, you need to assume that the knowledge of what a glowing cat signifies will also be preserved – perhaps through the preservation of historical records or the creation of some myth or legend. How could this be guaranteed?

Meanwhile, the semiotician Thomas Sebeok recommended the creation of a so-called 'Atomic Priesthood'. Members of the priesthood would preserve information about the waste repositories and hand it on to newly initiated members, ensuring a transfer of knowledge through the generations. Considered one way, this is not too different to our current system of atomic science, where a senior scientist passes on their knowledge to a PhD candidate. But still, putting such knowledge, and therefore power, into the hands of one small, elite group of people is a high-risk strategy easily open to abuse. It feels almost inevitable that at one point or other the knowledge will end up being used against the population it was meant to protect.

Perhaps a better way to warn our descendants about the waste is to talk to them not via atomic priests but directly, in the form of a message. At Andra's headquarters outside Paris, Dumont showed me a box. Inside, fixed in plastic cases, are two discs, each around 20 cm in diameter. The brainchild of

Dumont's predecessor, Patrick Charton, each disc is made of transparent industrial sapphire, onto which information is engraved using platinum.

With a mixture of regular and microscopic writing, a single disc could be engraved with the equivalent of 600 archive boxes or 400,000 sheets of A4 paper. On such a disc you could keep what Dumont calls a Key Information File, or KIF, around forty pages of important information telling future generations about the repository and the waste. 'And the last section is an international chapter, providing information about repositories all over the world, so it creates an international network,' Dumont said. Someone stumbling across Cigeo will also learn about Onkalo and all the other repositories yet to be built. Costing around €25,000 per disc, the sapphire (chosen for its durability and resistance to weathering and scratching) could last for nearly 2 million years – though one disc already has a crack in it, the result of a clumsy visitor at one of Andra's open days.

In the very long term, though, these plans also have a major drawback: how can we know that anyone living one million years in the future will understand any of the languages spoken today?

Think of the differences between modern and Old English. Without specialist education, who can understand the sentence 'Đunor cymð of hætan & of wætan'? That – meaning 'Thunder comes from heat and from moisture' – is only around 1,000 years old but already generally incomprehensible to the population of the British Isles.[8]

Languages also have a habit of disappearing. Around 4,000 years ago in the Indus Valley in what is now Pakistan and north-west India, for example, people were writing in a script that remains indecipherable to modern researchers.[9] In a million years it is unlikely that any language spoken today will still exist.

In the early 1990s, responding to another US government initiative, the architectural theorist Michael Brill partially side-stepped the issue of language by imagining a deterrent landscape 'non-natural, ominous, and repulsive', constructed of giant, menacing earthworks in the shape of jagged lightning bolts or other shapes that 'suggest danger to the body ... wounding forms, like thorns and spikes'. Anyone venturing further into the complex would then discover a series of standing stones with warning information about the radioactive waste written in many different languages and accompanied by basic pictograms – but even if these proved unreadable, the landscape itself should act as a warning. To help convey a sense of danger there would be carvings of human faces expressing horror and terror. One idea was to base them on Edvard Munch's *The Scream*.[10]

Florian Blanquer, a semiotician hired by Andra to work on a PhD about the problem of memory and radioactive waste, has studied this project. Brill's idea of using fear makes a lot of sense, he told me, 'because emotion is universal. Every single human on Earth, apart from those with some pathological issue, has a sense of fear, disgust and so on.' The drawback is that such a landscape – a strange, disturbing wonder – would probably attract rather than repel visitors. 'We are adventurers. We are drawn to conquer forbidding environments. Think about Antarctica, Mount Everest.' Or think about the twentieth-century European archaeologists, people not noticeably reticent when it came to opening up the tombs of Egyptian kings, no matter how many scary warnings and curses were inscribed on the walls.

With designs such as Brill's, our cultural fear of radiation was often part of the strategy for protecting the repositories. 'At the end of the last century the idea was to use fear or to deter people from digging or interacting with the site,' Dumont said. 'Now we are more into the idea of transmitting information and knowledge.'

One way that memory is transmitted is orally, from generation to generation. To study this, Dumont asked researchers to consider historical examples of oral transmission, using as a case study the construction and maintenance of the Canal du Midi between the Mediterranean and the Atlantic since the seventeenth century. Here, for over 300 years, the same families have worked on the project, passing down know-how from father to son. Dumont also talks about the need to ensure that as many people as possible hear about Cigeo. (Statistically, the more people that know about the repository, the greater the chances that the memory will be preserved.) As part of this strategy – in addition to, for example, hosting a visitor centre and giving interviews to journalists – for the past three years Andra has held a competition asking artists to suggest ways to mark the site. Models of some of these designs can be seen at the visitor centre. 'Artists provide new ideas and insights. They open our minds,' Dumont said. Les Nouveaux Voisins, winners of the 2016 prize, imagined constructing eighty 30-m-high concrete pillars, each with an oak tree planted at the top. As the years pass, the pillars will slowly sink and the oak trees replace them, leaving tangible traces both above and below the surface.

All of which is to say that working out how to communicate with the future is difficult. A friend of mine who once ran science workshops for children used to ask them to come up with ideas for signs for nuclear waste repositories. 'The kids hated this,' she wrote about the experience. 'We had to stop running the workshops after getting papers covered in red crayon scribbles of "Fuck Off." What does thousands of years even mean when you're ten years old and the weekend or even the end of the school day takes an eternity to arrive? The semioticians and linguists and archaeologists and material scientists started working on it in 1983, and they're still on the case. "Fuck Off" is a valid response.'

*

In north-east France I drove towards the village of Bure through a landscape patchworked with greens, from the russet of the woods to the bright limey green of a wheat field. A tiny village of limestone buildings centred round a town hall, Bure is the closest community to Cigeo and home to around ninety, mostly elderly, inhabitants. 'Young people can't stay here if they want to study and find jobs,' Benoît Jaquet told me. A village that once supported about ten farmers is now home to only two or three. Jaquet is the general secretary of the Comité Local d'Information et de Suivi (CLIS). Headquartered in the old communal laundry in the centre of the village, the CLIS is an organisation composed of local elected officials, representatives from trade unions and professional bodies, and environmental associations. Its purpose is to provide the local community with information about Cigeo, host public meetings and monitor the work of Andra by, for example, commissioning independent experts to review the agency's work.

If the repository is built, French law requires that the CLIS be transformed into a local commission that will last as long as the life of the repository. When members of the commission retire or leave public office, new members will take their place. 'So it's also a way to pass the baton,' Jaquet said. 'If there is a local commission there is a memory, [which is] not Andra's memory but an external memory.'

At the same time, Andra have set up three regional 'memory' groups, each composed of about twenty interested locals. They meet every six months and make their own suggestions for preserving and passing on the memory of the repository. Ideas so far include: collecting and preserving oral witness accounts; installing memory stones featuring themes and keywords related to the site; and developing an annual remembrance event to take place on the site, organised by and for the local people – a radioactive maypole ceremony or a nuclear 'beating the bounds'.

This last idea reverberates with the work of former NEA researchers Claudio Pescatore and Claire Mays, who have written: 'Do not hide these facilities; do not keep them apart, but make them a part of the community [...] something that belongs to the local, social fabric.' They go on to suggest that a monument celebrating the repository could be created, and argue that if it 'had a distinctiveness and aesthetic quality, would this not be one reason for communities to proudly own the site and maintain it?'[11]

Some members of the memory group favour this idea, but Jaquet, when I put it to him, seemed more sceptical. Could the repository, I asked, become one day a tourist destination? Something to attract visitors to the region?

On the contrary, he said. Some members of the CLIS 'say that every person living here will quit the district because of the risk, because of the image of the repository as a rubbish bin. Of course, some also think the repository will create employment and that this will become a new Silicon Valley. Maybe the reality will be somewhere between the two. But a tourist attraction? I'm not sure about that.'

Standing outside the headquarters of the CLIS, I watched a young, dreadlocked woman cross the small square in front of the town hall and enter a large, ramshackle stone house. Outside were several decrepit caravans and two oil drums mocked up as nuclear waste containers. A handmade signpost pointed towards Cigeo, the word 'Cigeo' vigorously crossed out. Over the main doorway was a banner: *Bure Zone Libre: La maison de résistance à la poubelle nucléaire* ('Free zone of Bure: house of resistance against nuclear waste').

Since 2004 this has been home to a revolving group of international anti-nuclear, anti-repository protestors. I thought about something Blanquer had told me: in his opinion, one of the most effective form of transmission had nothing to do with Andra at all – it was the existence of anti-repository groups such as the Maison de la Résistance.

By continually campaigning against Cigeo – and, presumably, by passing their beliefs on to their children – the protesters would necessarily keep the memory of the repository alive and in the public eye, their protests and clashes with the gendarmes reported in the news, the ramshackle stone house becoming its own sort of monument for Cigeo. 'So in fact the pro-repository groups need the anti-repository groups to stay alive in order to provide a good memory,' Blanquer said. 'Fortunately we are in France – in France there are always opponents to something!'

But there is one more radical proposal about how to deal with the threat of human intrusion and the preservation of memory: don't bother. Given our species's eternal curiosity, our capacity for violence and our historical inability to just leave well alone, might the safest thing to do with the repository be to hide it from our descendants? According to Jaquet: 'Some people here don't want a marker. Some are thinking that it would be better to forget ... It would be protection against terrorist action, for example.'

In Finland several people told me that because the repositories are passive systems, and because they will probably be buried far underground in areas with no deep natural resources, the question of memory preservation is moot. 'From above ground the site will be just like any other piece of forest or nature,' Jalonen said. There would be nothing to identify the repository, no reason for anyone to ever dig there. Out of sight, out of mind, seemed to be the idea. And after around, say, 100,000 years, almost all surface traces and any complex above-ground markers will vanish.[12] The only things left behind will be some slight indentations, perhaps a gentle protuberance or two. Things that to the untrained eye might appear to be only the 'natural' shape of the land. Eventually it will be as though no one had ever been there, as though there is nothing for anyone to remember.

Blanquer, however, warns that forgetting is not so easy: 'Forgetting is a passive action. You cannot say to yourself, "I will forget about that." It's like trying not to think about pink elephants. If you want to forget about [Cigeo], then first you have to get rid of any information about it. That would mean shutting down the web and destroying a lot of computers, a lot of newspapers, a lot of books.' In his opinion it is no longer possible for Cigeo to become, as the Danish film-maker Michael Madsen has said about Onkalo, the place *you must remember to forget*.[13]

*

Ultimately anyone wanting to preserve the memory of a repository will make use of a variety of strategies. Rely only on the transmission of knowledge between generations and you can never guarantee an unbroken line of succession. Rely only on direct communication and you risk leaving behind a message that, even if it survives physically, eventually no one will be able to understand.

To help them think about this problem, Andra asked Blanquer to conduct some research. If words are out, he concluded, you might use pictures to convey your message.

But many visual signs are still, like languages, culturally specific. Think about road signs, biohazard signs or the radiation trefoil. We only understand these because of prior cultural knowledge. Furthermore, we know that the meanings of signs are not always stable over time. The picture of a skull and crossbones, for instance, is usually connected either to piracy or to deadly toxic products. But, says Blanquer, 'for the medieval alchemists the skull represented Adam's skull and the crossed bones the cross that promised resurrection.' In just 600 or so years the sign has completely flipped from meaning life to meaning death.

Still, Blanquer thought that there was *one* universal sign – an image of a human being. 'If you go across America, across the UK, across Africa, across Europe, across Australia, you will see this sign and understand that men or women are being depicted.' [14] Furthermore, 'every human being [...] apprehends their body through space the same way as well. There is an up and down, a left and right, a front and back.' Pictographs (pictorial symbols for a word or phrase) based on an anthropomorphic figure in movement are likely to be recognised universally, he decided.

Now he had the beginnings of an idea, but it wasn't enough. You might draw a cartoon strip showing a person approaching a piece of radioactive waste, touching it and falling down.

But how can we guarantee that the panels will be read in the correct order? Or that touching the waste will be interpreted as a negative action? Read back to front it could look as though the waste had brought a dead man back to life. And how can a pictograph relying on the visual representation of tangible objects (such as a body) convey a message about radioactivity – something that can be neither seen nor touched?

In response to these problems, Blanquer has designed what he calls a 'praxeological device' – an 'integrated system that allows us to communicate a complex and abstract concept to all people at all times'.[15] Completely independent of any known language, the device works by teaching the person encountering it a brand-new language created especially for this purpose.

Blanquer envisages the device as a series of passages built underground, perhaps in the access tunnels of the repository. (Putting the device underground protects it from weathering and erosion.) On the wall of the first passage is a rectangular pictograph showing a person walking along the passage and a line of footprints indicating the direction of movement. (This relates to Blanquer's contention that the human body is a universal sign.) At the end of the corridor is a hole and a ladder and two more pictographs. A circular pictograph shows a person holding on to the ladder; a triangular pictograph shows a person not holding on and consequently falling off. At the bottom of the ladder is a second passage. Part-way along, the height of the roof suddenly drops. A circular pictograph shows a person ducking down and staying safe. A triangular pictograph shows a person not ducking and hitting their head.

In this way you begin to establish patterns: you learn, first, that the figure drawn on the walls relates to a person's actions within the device, and second, that you should copy the actions in the circles and avoid the actions in the triangles. And once these patterns are in place it should be possible to devise a way to deliver a more complex warning about the radioactive waste. Those final stages of the design are still in development, but one idea is to translate a tangible experience – such as the burns caused by fire – into a pictograph and then to use that as an analogy for radioactivity.

'What is really interesting is the idea of people learning by themselves,' Dumont said. 'Learning is important in the long term when you cannot just rely on transmission from generation to generation.'

*

Last summer I set out with some friends to walk part of the Ridgeway, an ancient long-distance route through the Chiltern Hills and North Wessex Downs. On Whiteleaf Hill in the Chilterns the chalky white path passes near the remains of a Neolithic barrow, around 5,000 years old.[16] You can tell immediately that it's not natural, the way the earth has been lumped up on the hillside, but today there is little to see except a low grassy mound with a view over the fields and woods of Buckinghamshire and the small town of Princes Risborough. We don't know who built the burial chamber or the name of the person interred there, what language they spoke and what they believed the world would be like in 5,000 years' time. Staring at the barrow, it was not continuity with the past but difference, distance, that I felt.

Sites such as the barrow on Whiteleaf Hill provided inspiration for Laure Boby, the winner of Andra's 2018 marker competition. She envisages three 5- to 10-m hills created from a mixture of local geological material (limestone, clay) and man-made material (concrete, plastics etc.). Just as we recognise the unnatural shape of the land at Whiteleaf, so any future archaeologists digging into the earth above Cigeo will realise immediately that something man-made is here, and that they should proceed with caution. 'Even if the memory was lost, the trace of the site would persist,' Boby has said. 'Just as we continue to observe – when walking or through Google Earth – traces of tumuli from ancient civilisations.' [17]

In the 1930s an archaeologist called Lindsay Scott broke open the Whiteleaf Hill barrow and discovered the remains of a human skeleton, about sixty pieces of pottery, flint shards and animal bones.[18] Assuming our species lasts long enough, our own civilisation will perhaps one day become as opaque, mysterious, unknowable as the Neolithic people who came before us. And just as we enter burial chambers in search of answers about the past, so archaeologists of the future may one day find themselves penetrating the concrete passageways and tunnels of Cigeo and Onkalo. Peering into the darkness they will ask themselves, 'Who built this place and what was its purpose? Is this a burial chamber? A military installation? The site of some lost religious ritual? Why did those long-dead people come here, digging down so far below the surface of the land? What were they running from? What were they trying to hide?'

*

On Olkiluoto, darkness fell by four o'clock. At the TVO visitor centre, hydrologist Anne Kontula and I sat drinking coffee. Working with the Finnish Meteorological Institute, Kontula is trying to determine Olkiluoto's future climate. Far beyond 5,000 years, at some point during the next 50,000 to 200,000 years the Earth will probably enter a new ice age.[19] Scientists have been studying what will happen to the canisters when ice up to 4 km thick exerts pressure on the bedrock, and how they would react if the temperature of the bedrock fell below freezing – in particular, whether the bentonite clay would retain its distinctive characteristics. 'One reason why we go 420 m with the depositories is that we are below the level at which we have had the permafrost in the previous ice ages,' Kontula told me.

If standing in the tunnels of Onkalo had been like stepping inside a giant memorial to the human species, it was also an expression of faith that the world will go on existing into the deep future, that there will still be life on planet Earth. 'One of the aspects of the work I like is that it has a bigger meaning,' Kontula said. 'It's about taking care of the dangerous waste that will be there even if we aren't.'

I sipped my coffee and stared across the water to where the lights of the power plants glistened in the distance. I was trying to imagine Olkiluoto when the ice finally comes. On the surface, everything will disappear. All the buildings, trees, grasses, rocks will vanish beneath the advancing glaciers.

'It's like when you're flying and you look down and see only clouds,' Kontula said, when I asked her. 'Everywhere you look it will be white. Just white.'

16

ON THE BEACH

At the end, I wanted to go back to the beginning. In Britain, the beginning is in far north-west Scotland. If you travel in a roughly diagonal line from the south-east to the north-west, from East Anglia, say, to the western Highlands, you travel back in time from the young Quaternary deposits of East Anglia, through the Palaeogene clays of the London Basin, the Upper Cretaceous chalk of the North Downs, the Jurassic oolite of the Cotswolds, the Triassic sandstones that ring the Pennines, the Carboniferous limestones of the Mendips, the Devonian Old Red Sandstone of the Brecon Beacons, the Ordovician and Silurian rocks of the Lake District, the Cambrian outcrops of the Central Grampians and up to the ancient rocks of the north-west Highlands and the Hebrides.

To find the very oldest rocks, I drove north from Inverness one rainy day in June. I was three months pregnant – which was like having a hangover all day without being able to drink – and eating Rich Tea biscuits against the nausea. Past Ullapool the wide road, sitting low and flat in the landscape, began to empty of traffic. On either side ash-coloured rocks erupted through the bracken and heather, and to the left a long, brown loch flashed in and out of sight. Habitation was sparse. In the distance I could see several hills that looked like children's drawings of hills, or a series of conical volcanoes with rounded tops.

The countryside I was driving through forms part of the UNESCO North West Highlands Geopark, an initiative to protect and promote outstanding geological sites. Looking out of the window I could see three distinct moments in deep time – the remains of three landscapes from three former worlds: Cambrian, Proterozoic and Archean.[1]

On the tops of the hills were flashes of white, which looked like snow but were in fact all that remains of a sequence of roughly 500-million-year-old quartzite from the Cambrian period, the first period of the Phanerozoic ('visible life') Eon and the moment when great numbers of complex life forms first appear in the fossil record. The hills themselves are Torridonian Sandstone, formed around 1 billion years ago from river sediments laid down after the erosion of a larger mountain range that once lay to the east, during the Proterozoic ('earlier life') Eon. Where the Pleistocene sands and gravels in the excavation on Cambridge Heath Road marked the move from human to deep time, here the sandstone marks the jump from the Phanerozoic (the eon that encompasses everything from 542 million years ago until the present day) into the Proterozoic – part of the Precambrian, the oldest, largest and most mysterious part of deep time.

The Precambrian sits at the bottom of the ICS chart – all livid purples and hot fuchsia pinks – subdivided into three Eons, the Proterozoic (earlier life), the Archean (beginning, origin) and, earliest of all, the Hadean (from Hades, god of the underworld and reflecting the presumably hellish conditions on Earth at that time). Together these Eons make up seven-eighths of all of deep time, but we know relatively little about them – and much of what we do know only began to come to light in the late twentieth and early twenty-first centuries. Constantly churning geological processes have left behind very little pristine material from the early Earth – the Torridonian Sandstones, for example, are unusual because few sedimentary rocks of that age survive intact. It wasn't until the 1960s, more than a hundred years after the discovery of the first dinosaur, that the field of Precambrian palaeobiology really took off.[2] Before then, many believed that there simply were no fossils in the Precambrian. And it was only in the last decades of the twentieth century that geologists began to identify a handful of Hadean rocks in western Greenland, north-western Canada and western Australia, including the 4-billion-year-old Acasta Gneiss I had seen auctioned at Christie's.

The third visible former world was the 'Cnoc and Lochan' landscape that the sandstone hills sit upon. In Gaelic, a *cnoc* is a small rock hill, while *lochan* are the small lakes often found in eroded hollows; the phrase describes the irregular, hummocky topography created by some of the oldest rocks in western Europe – the Lewisian Gneiss. These metamorphic rocks can be as much as 3 billion years old, placing them deep in the Archean Eon. They are the oldest rocks in Britain. The worlds they have persisted through encompass almost all the stories in this book.

*

The next day was warm and dry. I parked in a lay-by and followed a small stream uphill, away from the road. Underfoot was boggy with dark, reddish brown water. The path led between boulders of Lewisian Gneiss that broke through the grass and bracken, like the backs of grey whales in a green sea. I listed the rocks as I walked. Lewisian Gneiss. Torridonian Sandstone.

Cambrian Basal Quartzite. There's a pleasure in knowing the names of things. It's not about a need to categorise the world, sectioning it into little boxes. And clearly you don't have to know the names of rocks – or trees or plants or birds – in order to enjoy a landscape. But if you do have this information, something changes about the way you exist in that space. A named landscape thickens. It's to do with history and context but also, I think, with the quality of attention. To assign something its name, you need to take the time to pick out identifying features. You look for longer. And the more you know, the more things stop being a backdrop – blurred, indistinguishable, hurried over – and become somehow more present in the view, more insistently themselves, the way a familiar face stands out in a crowd.

I sat on a warm rock by the stream and ate a cheese sandwich. The water was cold, the colour of beer and swift-moving. A cuckoo was calling somewhere out of sight among the twisting silver birch trees that, further downhill, crowded round the stream. Up close, the gneiss I was sitting on was grey with crusty patches of white and pale fluorescent yellow lichen. If you spent a while staring at it, you could make out other colours in the grey: wavy lines of black, lines of salmon pink. In *The Geology of Britain* the geologist Peter Toghill describes this place as 'an exhumed Precambrian landscape'.[3] For millions of years it lay protected from the elements, buried underneath a continuous blanket of Torridonian Sandstone. Over time that sandstone wore away, leaving behind only those strangely shaped hills and revealing the preserved gneiss underneath. 'Had we stood here a billion years ago,' a BGS guide to the region explains, 'the contours of the Lewisian Gneiss would have been uncannily similar to those of today.'[4]

Uncannily similar but different. Scratch the grass, to begin with. Scratch the bracken and heather too. Scratch the silver birch trees and the cuckoo. Scratch the rich, peaty soil and leave behind a forbidding, rocky landscape, bare slopes piled with fans of scree, streams and rivers providing the only movement along, perhaps, with clouds of grit blown by the wind, and the rainfall drenching and darkening the rocks.

*

On the third day of my visit I drove down a long single-width road to the beach at Achmelvich. The water was the blue of Caribbean holiday advertisements. The sand was white, made from crushed shells. In the low, late afternoon sun, the gneiss glowed ash-grey with a blush of pale coral rising through, like the breast of a collared dove.

The scenery here has been compared to the glaciated shield landscapes of Canada and Sweden and New York's Central Park. It made me think of black-and-white Ingmar Bergman films set on islands and coastlines. After spending time in Edinburgh and the Scottish Lowlands, north-west Scotland looks like another country. But then, that's because it sort of is.

Around 600 million years ago, during the Precambrian, an ancient supercontinent situated around the South Pole began to break apart.[5] North-west Scotland went one way – along with North America, Canada and Greenland – while England, Wales and the rest of Scotland went another. This state persisted until around 420 million years ago – some 4 million years before the advent of the Devonian – when they began to come together again in the form the new supercontinent of Pangaea.

Over the next millions of years Scotland, England and Wales moved slowly from the edge to the centre of the new continent, becoming in turns warm and swampy and then dry, arid and fearsomely hot. Around 80 million years ago – the time of the chalk – Pangaea started dividing into the continental masses that we know today and the Atlantic Ocean began to form. Around 65 million years ago – just after the Chicxulub giant asteroid hit and possibly finished off the dinosaurs – Britain reached an approximation of its modern position.[6]

Scientists speculate that in 200–250 million years' time, as the continental plates continue to drift into the deep future of our planet, a new supercontinent will form.[7] There are competing theories as to what this will look like, but one version – named Novopangaea – has the Atlantic continuing to open, the Pacific continuing to close, until all the Earth's landmasses are joined together once more. Eurasia finds itself sandwiched between Africa on the west and North America on the east, with India, China and Australia pressing up from the south.

I walked around the headland at Achmelvich to a second bay and clambered down to another empty white beach. Here, the gneiss rose up above me to form a low cliff. The waves and the wind had cleaned and polished these rocks, displaying great bands of black and burnt orange (the orange colour was the result of iron-rich minerals). A jumble of fractured blocks, the cliff face was composed of so many lines and angles that when you stepped back it began to resemble a Cubist painting – a Braque, perhaps. Out in the bay, the sea had turned to dull silver cut with bright lozenges of light. I sat down with my back against the warm gneiss. In an early ultrasound at seven weeks we'd seen the baby: a tiny kidney-bean-shaped smudge with a white flickering heart. At twelve weeks we'd heard the heartbeat, fast and hurried as that of any small creature – a mouse, say, a field vole or darting shrew. The heart, I'd read, is the first organ to form during the building of the body. One of the earliest known hearts belonged to a 520-million-year-old arthropod from the Cambrian, when the white quartzite on the hilltops was laid down.[8]

Sitting on the beach, I could hear the regular heaving, shushing sound of the sea. Rocks and water: two substances on Earth that have always, more or less, resembled their modern forms. Where the bare rocks met the sea – if I turned my head so that I couldn't see any grass, any vegetation, any gulls swooping and diving – the world looked, I guessed, a little as it would have looked in the Precambrian, where everything on our planet has its origins. A simple and severe landscape, unadorned, unembellished, picked out mostly in shades of brown and grey. Shallow braided rivers sprawling over wide flood plains. The mineral smell of baking rocks. Torrents of glassy clear water pouring endlessly towards the early seas.

In the early Archean, before they became the Lewisian Gneiss, these rocks were predominantly igneous intrusions – grey to pink granitic rocks, dark grey base-rich gabbros – mixed with ancient sediments and a sequence of lavas. Deep within the Earth's crust they were buried in darkness, subjected to immense pressures and heat, and slowly transformed. By tracing the folds and faults, you can read the record of their turbulent journey towards the surface, where they finally broke through around 1,100 million years ago.

Once the rocks had escaped the Earth's crust, their form, give or take a little erosion, was fixed. Worlds slipped by them: landscapes of ice; hot, windswept deserts; silent, steaming swamps. Creatures came and went across their surface. Crawling, scuttling, running, leaping, walking. Through it all the rocks simply persisted, more or less unchanged, for over a billion years.

NOTES

1. Deep Time on Cambridge Heath Road

1. R. Feuda et al., 'Improved Modeling of Compositional Heterogeneity Supports Sponges as Sister to All Other Animals', *Current Biology* 27 (2017), p. 3864.

2. M. Bjornerud, 'Geology is Like Augmented Reality for the Planet', *Wired* (September 2018): https://www.wired.com/story/geology-is-like-augmented-reality-for-the-planet/

3. J. Morrison, 'The Blasphemous Geologist Who Rocked Our Understanding of Earth's Age', Smithsonian.com (August 2016): https://www.smithsonianmag.com/history/father-modern-geology-youve-never-heard-180960203/

4. S. Cotner, D. Brooks and R. Moore, 'Is the Age of the Earth One of Our "Sorriest Troubles?" Students' Perceptions about Deep Time Affect Their Acceptance of Evolutionary Theory', *Evolution* 64(3) (2010).

5. https://data.worldbank.org/indicator/SP.DYN.LE00.IN?locations=GB.

6. J. Playfair, *The Works of John Playfair, Esq.* (Edinburgh: Archibald Constable & Co., 1822), p. 81.

7. https://www.geolsoc.org.uk/history

8. C. Lyell, *Principles of Geology*, 7th edn. (London: John Murray, 1847), p. 190.

9. J. McPhee, *Annals of the Former World* (New York: Farrar, Straus and Giroux, 2000), p. 31.

10. N. Woodcock and R. Strachan (eds), *Geological History of Britain and Ireland* (Oxford: Blackwell Science Ltd, 2002), p. 4.

11. N. Woodcock and R. Strachan (eds), *Geological History of Britain and Ireland*, p. 4.

12. A. Tennyson, 'In Memoriam A. H. H.', in *Alfred Lord Tennyson: Selected Poems*, ed. C. Ricks (London: Penguin Classics 2007), p. 189.

2. Box 48903C16

1. https://unfccc.int/process-and-meetings/the-paris-agreement/the-paris-agreement

2. D. Carrington, 'Avoid Gulf Stream Disruption at All Costs, Scientists Warn', *The Guardian* (13 April 2018); D. J. R. Thornalley et al., 'Anomalously Weak Labrador Sea Convection and Atlantic Overturning during the Past 150 Years', *Nature* 556 (2018), pp. 227–30.

3. M. Walker et al., 'Formal Definition and Dating of the GSSP (Global Stratotype Section and Point) for the Base of the Holocene using the Greenland NGRIP Ice Core, and Selected Auxiliary Records', *Journal of Quaternary Science* 24 (2009), p. 3.

4. L. C. Sime et al., 'Impact of Abrupt Sea Ice Loss on Greenland Water Isotopes During the Last Glacial Period', *Proceedings of the National Academy of Sciences (PNAS)* 116 (2019), p. 4099.

5. X. Zhang et al., 'Abrupt North Atlantic Circulation Changes in Response to Gradual CO_2 Forcing in a Glacial Climate State', *Nature Geoscience* 10 (2017), p. 518.

6. E. Kintisch, 'The Great Greenland Meltdown', science.com (2017): https://www.sciencemag.org/news/2017/02/great-greenland-meltdown

7. L. D. Trusel et al., 'Nonlinear Rise in Greenland; and Runoff in Response to Post-Industrial Arctic Warming', *Nature* 564 (2018), p. 104.

8. A. Aschwanden 'The Worst is Yet to Come for the Greenland Ice Sheet', *Nature* 586 (2020), pp. 29–30.

3. Shallow Time

1. Quoted in D. B. McIntyre and A. McKirdy, *James Hutton: The Founder of Modern Geology* (Edinburgh: National Museums Scotland, 2012), p. 2.

2. McIntyre and McKirdy, *James Hutton*, p. 4.

3. M. J. S. Rudwick, *Earth's Deep History: How It Was Discovered and Why It Matters* (Chicago, IL: University of Chicago Press, 2014), p. 11.

4. S. Baxter, *Revolutions of the Earth: James Hutton and the True Age of the World* (London: Phoenix, 2004), p. 23.

5. J. Zalasiewicz, 'Encore des Buffonades, Mon Cher Count?', *The Paleontology Newsletter* 79 (2012), p. 4.

6. Zalasiewicz, 'Encore des Buffonades, Mon Cher Count?', p. 3.

7. Baxter, *Revolutions of the Earth*, p. 185.

8. Colin Campbell quoted by D. Cox, 'The Cliff That Changed Our Understanding of Time', bbc.com (2018): http://www.bbc.com/travel/story/20180312-how-siccar-point-changed-our-understanding-of-earth-history

9. Quoted in McIntyre and McKirdy, *James Hutton: The Founder of Modern Geology*, pp. 13–15.

10. Baxter, *Revolutions of the Earth*, p. 30.

11. McIntyre and McKirdy, *James Hutton: The Founder of Modern Geology*, p. 13.

12. Baxter, *Revolutions of the Earth*, pp. 96–7.

13. Baxter, *Revolutions of the Earth*, p. 93.

14. Quoted in McIntyre and McKirdy, *James Hutton: The Founder of Modern Geology*, p. 16.

15. *Deep Time*, episode 1 (BBC Two, 2010): https://www.bbc.co.uk/programmes/b00wkc23

16. Quoted in P. Lyle, *The Abyss of Time: A Study in Geological Time and Earth's History* (Edinburgh: Dunedin Academic Press, 2016), p. 25.

17. McIntyre and McKirdy, *James Hutton: The Founder of Modern Geology*, p. 34.

18. McIntyre and McKirdy, *James Hutton: The Founder of Modern Geology*, p. 19.
19. Quoted in Lyle, *The Abyss of Time*, p. 50.
20. Baxter, *Revolutions of the Earth*, p. 185.
21. McIntyre and McKirdy, *James Hutton: The Founder of Modern Geology*, p. 16.
22. E. Kolbert, *The Sixth Extinction: An Unnatural History* (London: Bloomsbury, 2014), p. 50.
23. C. Darwin, *The Works of Charles Darwin*, vol. 15, *On the Origin of Species* (New York: New York University Press, 1988), p. 202.
24. R. Fortey, 'Charles Lyell and Deep Time', *Geoscientist* 21(9) (2011): https://www.geolsoc.org.uk/Geoscientist/Archive/October-2011/Charles-Lyell-and-deep-time
25. Lyell, *Principles of Geology*, p. 166.

4. The Auctioneer

1. V. F. Buchwald, *Handbook of Iron Meteorites* (Berkeley, CA: University of California Press, 1977), p. 1123.
2. http://meteorites.wustl.edu/rlk.htm
3. http://adsabs.harvard.edu/full/1998ncdb.conf...33S
4. https://atlas.fallingstar.com/home.php
5. http://curious.astro.cornell.edu/about-us/75-our-solar-system/comets-meteors-and-asteroids/meteorites/313-how-many-meteorites-hit-earth-each-year-intermediate
6. https://www.nasa.gov/mission_pages/asteroids/overview/fastfacts.html
7. https://www.livescience.com/36981-ancient-egyptian-jewelry-made-from-meteorite.html
8. https://www.livescience.com/36981-ancient-egyptian-jewelry-made-from-meteorite.html
9. J. Nobel, 'The True Story of History's Only Known Meteorite Victim', *National Geographic News* (2013): https://www.nationalgeographic.com/

news/2013/2/130220-russia-meteorite-ann-hodges-science-space-hit/

10. Nobel, 'The True Story of History's Only Known Meteorite Victim'.

5. **Time Lords**

1. M. Walker et al., 'Formal Ratification of the Subdivision of the Holocene Series/Epoch (Quaternary System/Period): Two New Global Boundary Stratotype Sections and Points (GSSPs) and Three New Stages/Subseries', *Episodes* 41(4) (2018), p. 213.

2. R. Meyer 'Geology's Timekeeper's Are Feuding', *The Atlantic* (2018): https://www.theatlantic.com/science/archive/2018/07/anthropocene-holocene-geology-drama/565628/

3. S. Lewis and M. Maslin, 'Anthropocene vs. Meghalayan: Why Geologists Are Fighting over Whether Humans Are a Force of Nature', *The Conversation* (2018): https://theconversation.com/anthropocene-vs-meghalayan-why-geologists-are-fighting-over-whether-humans-are-a-force-of-nature-101057

4. S. P. Hesselbo et al., 'Massive Dissociation of Gas Hydrate during a Jurassic Oceanic Anoxic Event', *Nature* 406 (2000), pp. 392–5.

5. G. Dera and Y. Donnadieu, 'Modeling Evidences for Global Warming, Arctic Seawater Freshening, and Sluggish Oceanic Circulation during the Early Toarcian Anoxic Event', *Paleoceanography and Paleoclimatology* 27(2) (2012).

6. http://www.stratigraphy.org/index.php/ics-chart-timescale

7. M. O. Clarkson et al., 'Ocean Acidification and the Permo-Triassic Mass Extinction', *Science* 348 (2016).

8. E. Vaccari, 'The "Classification" of Mountains in Eighteenth Century Italy and the Lithostratigraphic Theory of Giovanni Arduino (1714–1795)', *Geology Society of America*, special paper 411 (2006), p. 157.

9. http://palaeo.gly.bris.ac.uk/Russia/Russia-Murchison.html

10. F. M. Gradstein et al., 'Chronostratigraphy: Linking Time and

Rock', in F. M. Gradstein, J. G. Ogg and A. G. Smith (eds), *A Geologic Time Scale 2004* (Cambridge: Cambridge University Press, 2004), p. 21.

11. Gradstein et al., 'Chronostratigraphy: Linking Time and Rock', p. 21.

12. M. J. Head and P. L. Gibbard, 'Formal Subdivision of the Quaternary System/Period: Past, Present, and Future', *Quaternary International* (2015), p. 1040.

13. J. Rong, 'Report of the Restudy of the Defined Global Stratotype of the Base of the Silurian System', *Episodes* 31(3) (2008), pp. 315–18.

14. Walker et al., 'Formal Ratification of the Subdivision of the Holocene Series/Epoch (Quaternary System/Period)', p. 213.

15. P. J. Crutzen and E. F. Stoermer, 'The "Anthropocene"', *Global Change Newsletter* 41 (2000), p. 17.

16. Lewis and Maslin, 'Anthropocene vs Meghalayan'.

17. Ibid.

18. P. Voosen, 'Massive Drought or Myth? Scientists Spar over an Ancient Climate Event behind Our New Geological Age', sciencemag.org (2018): https://www.sciencemag.org/news/2018/08/massive-drought-or-myth-scientists-spar-over-ancient-climate-event-behind-our-new

19. Ibid.

6. The Demon in the Hills

1. https://www.geolsoc.org.uk/Plate-Tectonics/Chap2-What-is-a-Plate

2. J. McPhee, *Annals of the Former World* (New York: Farrar, Straus and Giroux, 2000), p. 148.

3. F. J. Vine and D. H. Matthews, 'Magnetic Anomalies over Oceanic Ridges', *Nature*, 199 (1963), pp. 947–9.

4. W. C. Pitman III and J. R. Heirtzler, 'Magnetic Anomalies over the Pacific-Antarctic Ridge', *Science* 154 (1966), pp. 1164–71.

5. 'The North Pacific: An Example of Tectonics on a Sphere', *Nature* 216 (1967), pp. 1276–80.

6. X. Le Pichon, 'Sea-Floor Spreading and Continental Drift', *Journal of Geophysical Research* 73(12) (1968), pp. 3661–97.

7. A. Wegener, *The Origin of Continents and Oceans,* trans. J. Biram (Mineola, NY: Dover Publications, 2003).

8. Quoted in R. Conniff, smithsonianmag.com (2012): https://www.smithsonianmag.com/science-nature/when-continental-drift-was-considered-pseudoscience-90353214/

9. T. Atwater, 'When the Plate Tectonics Revolution Met Western North America', in N. Oreskes (ed.), *Plate Tectonics, An Insider's History of the Modern Theory of the Earth*, ed. (Boulder, CO: Westview Press, 2001), pp. 243–63.

10. W. J. Morgan, 'Rises, Trenches, Great Faults, and Crustal Blocks', *Journal of Geophysical Research* 73(6) (1968), pp. 1959–82.

11. https://pubs.usgs.gov/gip/earthq3/safaultgip.html

12. https://pubs.usgs.gov/fs/2015/3009/pdf/fs2015–3009.pdf

13. https://pubs.usgs.gov/fs/2015/3009/pdf/fs2015–3009.pdf

14. Quoted in H. E. Le Grand, 'Plate Tectonics, Terranes and Continental Geology', in D. R. Oldroyd (ed.), *The Earth Inside and Out*, Geological Society Special Publication 192 (Bath: Geological Society, 2002), p. 202.

15. T. Atwater, 'Implications of Plate Tectonics for the Cenozoic Tectonic Evolution of Western North America', *GSA Bulletin* 81(12) (1970), pp. 3513–36.

16. https://www.usgs.gov/faqs/will-california-eventually-fall-ocean?qt-news_science_products=0#qt-news_science_products

17. Oreskes (ed.), *Plate Tectonics*.

18. Quoted in Natalie Angier 'Plate Tectonics May Be Responsible for Evolution of Life on Earth, Say Scientists', *The Independent* (2019): https://www.independent.co.uk/environment/earth-shell-cracked-global-warming-tectonic-plates-mantle-geology-science-a8690606.html

19. https://www.usgs.gov/faqs/can-you-predict-earthquakes?qt-news_science_products=0#qt-news_science_products

20. https://leginfo.legislature.ca.gov/faces/codes_displayText.xhtml?division=2.&chapter=7.5.&lawCode=PRC

21. L. M. Jones et al., *The ShakeOut Scenario,* US Department of Interior/US Geological Survey (2008).

22. Jones et al., *The ShakeOut Scenario*, pp. 9, 6, 10.

23. S. E. Hough, *Finding Fault in California: An Earthquake Tourist's Guide* (Missoula, MO: Mountain Press, 2004), p. 35.

24. D. L. Ulin, *The Myth of Solid Ground* (New York: Penguin Books, 2005), p. 8.

25. Hough, *Finding Fault in California*, p. 42.

26. Hough, *Finding Fault in California*, p. 44.

27. Hough, *Finding Fault in California*, p. 201.

28. https://www.usgs.gov/faqs/can-you-predict-earthquakes?qt-news_science_products=0#qt-news_science_products

29. S. E. Hough, *Predicting the Unpredictable: The Tumultuous Science of Earthquake Prediction* (Princeton, NJ: Princeton University Press, 2010), p. 96.

30. Hough, *Predicting the Unpredictable*, p. 84.

31. C. King, https://thecharlottekingeffect.com/page/3/

32. C. King, https://thecharlottekingeffect.com/about/

33. C. King, https://thecharlottekingeffect.com/page/3/

34. Quoted in Ulin, *The Myth of Solid Ground*, p. 36.

35. Quoted in Hough, *Predicting the Unpredictable*, p. 166.

36. Ulin, *The Myth of Solid Ground*, pp. 34–73.

37. S. J. Gould , 'The Rule of Five', *The Flamingo's Smile* (New York: W. W. Norton, 1985), p. 199.

38. Hough, *Predicting the Unpredictable*, p. 222.

7. A Lost Ocean

1. F. Pryor, *The Making of the British Landscape* (London: Allen Lane, 2010), p. 138.

2. https://www.bgs.ac.uk/about/ourPast.html

3. Office for National Statistics, 1911 Census General Report with Appendices; Office for National Statistics (1917), 2011 UK Census aggregate data, UK Data Service (2016).

4. G. Gohau, rev. and trans. A.V. Carozzi and M. Carozzi, 'The Use of Fossils', in *A History of Geology* (New Brunswick, NJ: Rutgers University Press, 1990), pp. 136–7.

5. Gohau, 'The Use of Fossils', pp. 136–7.

6. https://www.geolsoc.org.uk/Library-and-Information-Services/ Exhibitions/William-Strata-Smith/Stratigraphical-theories

7. *Proceedings of the Geological Society* 1 (1831), p. 271.

8. http://www.strata-smith.com/?page_id=279

9. G. White, *The Natural History of Selborne* (London: Penguin Classics, 1977), p. 145.

10. R. Kipling, 'Sussex', *The Cambridge Edition of the Poems of Rudyard Kipling,* ed. T. Pinney (Cambridge: Cambridge University Press, 2013).

11. R. C. Selley, *The Winelands of Britain: Past, Present and Prospective* (Dorking: Petravin Press, 2008).

12. D. T. Aldiss et al., 'Geological Mapping of the Late Cretaceous Chalk Group of Southern England: A Specialised Application of Landform Interpretation', *Proceedings of the Geologists' Association* 123(5) (2015), pp. 728–41.

13. Quoted in K. Smale, 'Bricks Sent Flying During Crossrail Tunnelling', *New Civil Engineer* (2018): https://www. newcivilengineer.com/latest/revealed-bricks-sent-flying-during-crossrail-tunnelling-08–10–2018

14. https://www.geolsoc.org.uk/GeositesChannelTunnel

15. M. A. Woods, 'Applied Palaeontology in the Chalk Group: Quality Control for Geological Mapping and Modelling and Revealing New Understanding', *Proceedings of the Geologists' Association* 126 (2015), pp. 777–87.

16. Quoted in P. Laity, 'Eric Ravilious: Ups and Downs', *The Guardian* (30 April 2011).

17. C. D. Clark et al., 'Pattern and Timing of Retreat of the Last

British-Irish Ice Sheet', *Quaternary Science Reviews* 44 (2012), p. 112.

8. The Fiery Fields

1. W. Hamilton, *Observations on Mount Vesuvius, Mount Etna, and Other Volcanoes* (London: T. Cadell, 1774), pp. 128–32.
2. T. Ricci et al., 'Volcanic Risk Perception in the Campi Flegrei Area', *Journal of Volcanology and Geothermal Research* (2013), p. 124.
3. http://volcanology.geol.ucsb.edu/pliny.htm
4. http://www.ov.ingv.it/ov/en.html
5. D. Hunter, 'The Cataclysm: "Vancouver! Vancouver! This Is It!"' (2012): https://blogs.scientificamerican.com/rosetta-stones/the-cataclysm-vancouver-vancouver-this-is-it/
6. G. Chiodini et al., 'Magma near the Critical Degassing Pressure Drive Volcanic Unrest towards a Critical State', *Nature Communications* 7 (2016), p. 13712.
7. C. R. J. Kilburn et al., 'Progressive Approach to Eruption at Campi Flegrei Caldera in Southern Italy', *Nature Communications* 8 (2017), p. 15312.
8. S. de Vita et al., 'The Agnano-Spina Eruption (4100 years BP) in the Restless Campi Flegrei Caldera (Italy)', *Journal of Volcanology and Geothermal Research* 91 (1999), p. 269.
9. http://www.pacificdisaster.net/pdnadmin/data/original/JB_DM311_PNG_1994_disaster_management.pdf

9. Ammonite

1. https://www.nhm.ac.uk/discover/mary-anning-unsung-hero.html;
2. H. Torrens, 'Mary Anning (1799–1847) of Lyme; "the greatest fossilist the world ever knew"', *British Journal for the History of Science (BJHS)* 28 (1995), p. 258.
3. A. Singh, 'Film-Makers Create Fictional Same-Sex Romance

To Spice Up Story of "unsung hero of fossil world"', *The Telegraph* (11 March 2019).

4. M. Doody, *Jane Austen's Names* (Chicago, IL: University of Chicago Press, 2015), pp. 367–8.

5. https://www.nhm.ac.uk/discover/mary-anning-unsung-hero.html

6. Torrens, 'Mary Anning (1799–1847) of Lyme', p. 260.

7. https://www.dorsetecho.co.uk/news/9628097.1yme-regis-residents-delighted-by-195m-project-to-save-homes/

8. Torrens, 'Mary Anning (1799–1847) of Lyme', p. 269.

9. Torrens, 'Mary Anning (1799–1847) of Lyme', p. 257.

10. J. Zalasiewicz, 'The Very Dickens of a Palaeontologist', *The Paleontology Newsletter* 80 (2012), p. 4.

11. Torrens, 'Mary Anning (1799–1847) of Lyme', p. 265.

12. Zalasiewicz, 'The Very Dickens of a Palaeontologist', p. 3.

13. Zalasiewicz, 'The Very Dickens of a Palaeontologist', p. 7.

14. Quoted in B. Chambers, 'Mary Anning: Fossil Hunter', in S. Charman-Anderson (ed.), *More Passion for Science: Journeys into the Unknown* (London: Finding Ada, 2015).

15. https://www.theuniguide.co.uk/subjects/geology/ https://eos.org/agu-news/working-toward-gender-parity-in-the-geosciences

16. HESA

17. https://www.americangeosciences.org/geoscience-currents/ female-geoscience-faculty-representation-grew-steadily-between-2006-2016/

18. https://www.wisecampaign.org.uk/statistics/ annual-core-stem-stats-round-up-2019-20/

19. https://www.aauw.org/resources/research/the-stem-gap/

20. J. Zalasiewicz et al., 'Scale and Diversity of the Physical Technosphere: A Geological Perspective', *The Anthropocene Review* 4(1) (2017), p. 10.

21. T. Hardy, *A Pair of Blue Eyes* (Ware: Wordsworth Classics, 1995), p. 172.

22. *The Quarterly Journal of the Geological Society of London* 4 (1848), p. 24.

23. Torrens, 'Mary Anning (1799–1847) of Lyme', p. 257.

24. https://trowelblazers.com/

25. Torrens, 'Mary Anning (1799–1847) of Lyme', p. 269.

10. The First Forest

1. W. E. Stein et al., 'Surprisingly Complex Community Discovered in the Mid-Devonian Fossil Forest at Gilboa', *Nature* 483 (2012), p. 78.

2. L. VanAller Hernick, *The Gilboa Fossils* (New York: New York State Museum, 2003), p. 1.

3. VanAller Hernick, *The Gilboa Fossils*, p. 3.

4. VanAller Hernick, *The Gilboa Fossils*, p. 4.

5. C. M. Berry, The Rise of Earth's Early Forests, *Cell Biology* 29(16) (2019), pp. 792–794.

6. P. Giesen and C. M. Berry 'Reconstruction and Growth of the Early Tree Calamophyton (Pseudosporochnales, Cladoxylopsida) Based on Exceptionally Complete Specimens from Lindlar, Germany (Mid-Devonian)', *International Journal of Plant Science* 174(4) (2013), pp. 665–86.

7. C. M. Berry et al., 'Unique Growth Strategy in the Earth's First Trees Revealed in Silicified Fossil Trunks from China', *PNAS* 114(45) (2017), p. 12009.

8. Berry et al., 'Unique Growth Strategy', p. 12009.

9. C. M. Berry, 'How the First Trees Grew So Tall with Hollow Cores – New Research', *The Conversation* (23 October 2017)

10. VanAller Hernick, *The Gilboa Fossils*, p. 23.

11. VanAller Hernick, *The Gilboa Fossils*, p. 37.

12. C. M. Berry, The Rise of Earth's Early Forests, *Cell Biology* 29(16) (2019), pp. 792–794.

13. C. M. Berry and J. E. Marshall, 'Lycopsid Forests in the Early Late Devonian Paleoequatorial Zone of Svalbard', *Geology*, 43(12) (2015), pp. 1043–6.

14. Stein et al., 'Surprisingly Complex Community Discovered', p. 79.

15. https://www.sciencedirect.com/science/article/abs/pii/ S0960982219315696

16. https://www.cell.com/current-biology/fulltext/S0960-9822(19)30861-9?_returnURL=https%3A%2F%2Flinkinghub. elsevier.com%2Fretrieve%2Fpii%2FS0960982219308619%3Fsh owall%3Dtrue.

17. T. Algeo and S. E. Scheckler, 'Terrestrial-Marine Teleconnections in the Devonian: Links between the Evolution of Land Plants, Weathering Processes, and Marine Anoxic Events', *Philosophical Transactions of the Royal Society, London B* 353 (1998), pp. 113–30.

11. What We Talk About When We Talk About Dinosaurs

1. https://www.npr.org/2018/07/10/627782777/many-paleontologists-today-are-part-of-the-jurassic-park-generation

2. D. Naish and P. M. Barrett, *Dinosaurs: How They Lived and Evolved* (London: Natural History Museum, 2018), p. 4.

3. G. Mantell, 'Notice on the Iguanodon, a Newly Discovered Fossil Reptile, from the Sandstone of Tilgate, in Sussex', *Philosophical Transactions of the Royal Society* 115 (1825), pp. 179–86.

4. G. Mantell, *The Geology of the South East of England* (London: Longman, 1833), p. 318.

5. W. Buckland, 'Notice on the Megalosaurus or Great Fossil Lizard of Stonesfield', *Transactions of the Geological Society of London* (2)1 (1824), pp. 390–96.

6. Naish and Barrett, *Dinosaurs*, p. 14.

7. Naish and Barrett *Dinosaurs*, p. 17.

8. Naish and Barrett, *Dinosaurs*, pp. 18–20.

9. Quoted in R. Black, smithsonian.com (16 November 2009): https://www.smithsonianmag.com/science-nature/ jingo-the-dinosaur-a-world-war-i-mascot-57348765/

10. W. L. Stokes, *The Cleveland-Lloyd Dinosaur Quarry: Window to the Past* (Washington DC: US Department of the Interior, Bureau of Land Management, 1985).

11. Naish and Barrett, *Dinosaurs*, pp. 20–22.

12. https://www.nhm.ac.uk/discover/dino-directory/allosaurus.html

13. Quoted in R. Black, smithsonianmag.com (10 July 2015): https://www.smithsonianmag.com/science-nature/what-killed-dinosaurs-utahs-giant-jurassic-death-pit-180955878/

14. Ibid.

15. K. Carpenter, 'Evidence for Predator–Prey Relationships', in K. Carpenter (ed.), *The Carnivorous Dinosaurs* (Bloomington, IN: Indiana University Press, 2005), p. 332.

16. M. Reynolds, 'The Dinosaur Trade', wired.co.uk (21 June 2018): https://www.wired.co.uk/article/dinosaur-t-rex-auction-sale-private-fossil-trade

17. Ibid.

18. http://vertpaleo.org/GlobalPDFS/SVP-to-Aguttes-about-Theropod,-2018-english.aspx

19. J. Pickrell, 'Carnivorous-Fossil Auction Reflects Rise in Private Fossil Sales', nature.com (1 June 2018): https://www.nature.com/articles/d41586-018-05299-3

20. Naish and Barrett, *Dinosaurs*, p. 204.

21. R. J. Whittle et al., 'Nature and Timing of Biotic Recovery in Antarctic Benthic Marine Ecosystems Following the Cretaceous–Palaeogene Mass Extinction', *Palaeontology* 62(6) (2019), p. 919.

22. J. Zalasiewicz *The Earth after Us: What Legacy Will Humans Leave in the Rocks?* (Oxford: Oxford University Press, 2008), pp. 191–2.

23. Naish and Barrett, *Dinosaurs*, pp. 5–6.

12. Colouring Deep Time

1. J. Vinther, 'The True Colours of Dinosaurs', *Scientific American* 16(3) (2017), p. 52.

2. J. Vinther, 'A Guide to the Field of Palaeo Colour', *Bioessays* 37 (2015), pp. 643–56.

3. J. Vinther, 'Fossil Melanosomes or Bacteria? A Wealth of Findings Favours Melanosomes', *Bioessays* 38 (2015), p. 220.

4. J. Vinther et al., 'The Colour of Fossil Feathers', *Biology Letters* (2008) vol. 4, pp. 522–5.

5. Q. Li et al., 'Plumage Colour Patterns of an Extinct Dinosaur', *Science* 327(5971) (2010), pp. 1369–72; F. Zhang et al., 'The Colour of Cretaceous Dinosaurs and Birds', *Nature* 463 (7282) (2010), pp. 1075–8.

6. J. Lindgren et al., 'Interpreting Melanin-Based Coloration through Deep Time: A Critical Review', *Proceedings of the Royal Society* (2015).

7. J. Hawkes, *A Land* (Boston, MA: Beacon Press, 1991), p. 77.

8. M. E. McNamara et al., 'The Fossil Record of Insect Colour Illuminated by Maturation Experiments', *Geology* 41(4) (2013), pp. 487–90.

9. M. E. McNamara et al., 'Reconstructing Carotenoid-Based and Structural Coloration in Fossil Skin', *Current Biology* 26 (2016), pp. 1–8.

10. M. E. McNamara, 'The Taphonomy of Colour in Fossil Insects and Feathers', *Palaeontology* 56(3) (2013), pp. 557–75.

11. A. Dance, 'Prehistoric Animals in Living Colour', *PNAS* 113(31) (2016), pp. 8552–6.

12. Q. Li et al., 'Reconstruction of *Microraptor* and the Evolution of Iridescent Plumage', *Science* 335 (2012), pp. 1215–19.

13. J. Vinther et al., '3D Camouflage in an Ornithischian Dinosaur', *Current Biology* 26(18) (2016), pp. 2456–62.

14. Naish and Barrett, *Dinosaurs*.

13. Urban Geology

1. R. Siddall, 'Rome in London: The Marbles of the Brompton Oratory', *Urban Geology in London* 28 (2015), http://www.ucl.ac.uk/~ucfbrxs/Homepage/walks/Brompton.pdf
2. R. Caillois, trans. B. Bray, *The Writing of Stones* (Charlottesville, VA: University of Virginia Press, 1988).
3. https://geologistsassociation.org.uk/about/
4. T. Nield, *Underlands: A Journey through Britain's Lost Landscape* (London: Granta, 2014), p. 145.
5. T. Hardy, *A Pair of Blue Eyes* (Ware: Wordsworth Classics, 1995), p. 172.
6. V. Morra et al., 'Urban Geology: Relationships between Geological Setting and Architectural Heritage of the Neapolitan Area', *Journal of the Virtual Explorer* 36 (2010).
7. Ibid.
8. https://ec.europa.eu/regional_policy/en/projects/major/italy/major-redevelopment-of-naples-historic-centre

14. In Search of the Anthropocene

1. J. McPhee, 'Basin and Range', *Annals of the Former World* (New York: Farrar, Straus and Giroux, 2000), p. 90.
2. A. Ganopolski et al., 'Critical Isolation – CO_2 Relation for Diagnosing Past and Future Glacial Inception', *Nature* 529 (2016), pp. 200–03.
3. C. Waters et al., 'The Anthropocene is Functionally and Stratigraphically Distinct from the Holocene', *Science* 351 (2016), p. 8.
4. Waters et al., 'The Anthropocene', p. 8.
5. J. Zalasiewicz et al., 'Petrifying Earth Process', *Theory, Culture and Society* 34 (2017), pp. 83–104.
6. J. Zalasiewicz et al., 'Human Bioturbation, and the Subterranean Landscape of the Anthropocene', *Anthropocene* 6 (2014), pp. 3–9.
7. Zalasiewicz et al., 'Human Bioturbation', p. 3.

8. J. Zalasiewicz et al., 'Scale and Diversity of the Physical Technosphere: A Geological Perspective', *The Anthropocene Review* (2016), pp. 1–14.

9. Zalasiewicz et al., 'Scale and Diversity', p. 11.

10. J. Zalasiewicz et al., 'The Working Group on the Anthropocene: Summary of Evidence and Interim Recommendations', *Anthropocene* 19 (2017), pp. 55–60.

11. M. Maslin and S. Lewis, 'Defining the Anthropocene', *Nature* 519(7542) (2015), p. 171.

12. Waters et al., 'The Anthropocene', p. 8.

13. S. C. Finney and L. E. Edwards, 'The "Anthropocene" Epoch: Scientific Decision or Political Statement?', *GSA Today* 26(3) (2016), p. 9.

14. S. C. Finney, 'The 'Anthropocene' as Ratified Unit in the ICS International Chronostratigraphic Chart: Fundamental Issues That Must Be Addressed by the Task Group', in C. N. Waters et al. (eds), *A Stratigraphical Basis for the Anthropocene*, Geological Society special publication 395 (London: Geological Society, 2014), p. 27.

15. P. L. Gibbard and M. J. C. Walker, 'The Term "Anthropocene" in the Context of Formal Geological Classification', in Waters et al. (eds), *A Stratigraphical Basis for the Anthropocene*, pp. 29–37.

16. J. Zalasiewicz et al., 'Making the Case for a Formal Anthropocene Epoch: An Analysis of Ongoing Critiques', *Newsletter on Stratigraphy* 50(2) (2017), p. 207.

17. E. W. Wolff, 'Ice Sheets and the Anthropocene', Geological Society special publications 395 (2013), pp. 255–63.

18. Zalasiewicz et al., 'Making the Case for a Formal Anthropocene Epoch', pp. 208–9.

15. 'This Place is Not a Place of Honor'

1. https://www.gov.uk/government/publications/

ionising-radiation-dose-comparisons/
ionising-radiation-dose-comparisons

2. https://www.arpansa.gov.au/understanding-radiation/what-is-radiation/ionising-radiation/beta-particles; https://www.gov.uk/government/publications/ionising-radiation-dose-comparisons/ionising-radiation-dose-comparisons

3. R. C. Ewing et al., 'Geological Disposal of Nuclear Waste: A Primer', *Elements* 12(4) (2016), pp. 233–7.

4. http://www.posiva.fi/en/final_disposal/basics_of_the_final_disposal#.XfiuS5P7TOQ

5. https://www.world-nuclear.org/information-library/safety-and-security/safety-of-plants/chernobyl-accident.aspx

6. Human Interference Task Force, 'Reducing the Likelihood of Future Human Activities That Could Affect Geologic High-Level Waste Repositories' (Columbus, OH: Office of Nuclear Waste Isolation, 1984).

7. F. Blanquer, 'Building Sustainable and Efficient Markers to Bridge Ten Millennia', *44th Annual Waste Management Conference (WM2018)* (Tempe, AZ: Waste Management Symposia, Inc., 2018), p. 5701.

8. https://public.oed.com/blog/old-english-an-overview/

9. A. Robinson, 'Ancient Civilization: Cracking the Indus Script', *Nature* 526 (2015), pp. 499–501.

10. K. M. Trauth et al., *Expert Judgement on Markers to Deter Inadvertant Human Intrusion into the Waste Isolation Pilot Plant* (Albuquerque, NM: Sandia National Laboratories, 1993).

11. C. Pescatore and C. Mays, *Records, Marks and People: For the Safe Disposal of Radioactive Waste* (Stockholm: Swedish Nuclear Power Inspectorate, 2009): https://www.osti.gov/etdeweb/biblio/971770

12. D. Harmand and J. Brulhet, 'Could the Landscape Preserve Traces of a Deep Underground Nuclear Waste Repository over the Very Long Term? What We Can Learn from the Archaeology of Ancient Mines', *Radioactive Waste*

*Management and Constructing Memory for Future
Generations. Proceedings of the International Conference and
Debate, 15–17 September 2014, Verdun, France* (2015).
13. M. Madsen (dir.), *Into Eternity*, prod. Lise Lense-Møller
(2010).
14. Blanquer, 'Building Sustainable and Efficient Markers', p. 5701.
15. Ibid.
16. https://historicengland.org.uk/listing/the-list/list-entry/1009532
17. L. Boby, https://www.andra.fr/sites/default/files/2019–03/
ArtEtMemoire2019-Termen%204.pdf
18. https://historicengland.org.uk/listing/the-list/list-entry/1009532
19. Ganopolski et al., 'Critical Isolation', pp. 200–03.

16. On the Beach
1. J. Mendum, J. Merritt and A. McKirdy, *Northwest Highlands:
A Landscape Fashioned by Geology* (Perth: Scottish Natural
Heritage, 2001).
2. J. William Schopf, 'Solution to Darwin's Dilemma: Discovery
of the Missing Precambrian Record of Life', *PNAS* 97(13)
(2000), p. 6947.
3. P. Toghill, 'Britain during the Precambrian', *The Geology of
Britain* (Ramsbury: Crowood Press, 2002), p. 23.
4. Mendum, Merritt and McKirdy, *Northwest Highlands*, p. 13.
5. Mendum, Merritt and McKirdy, *Northwest Highlands*, p. 6.
6. Mendum, Merritt and McKirdy, *Northwest Highlands*, p. 7.
7. M. Green et al., 'What Planet Earth Might Look Like When
the Next Supercontinent Forms – Four Scenarios', *The
Conversation* (November 2018): https://theconversation.
com/what-planet-earth-might-look-like-when-the-next-
supercontinent-forms-four-scenarios-107454
8. X. Ma et al., 'An Exceptionally Preserved Arthropod
Cardiovascular System from the Early Cambrian', *Nature
Communications* 5 (2014); H. Dunning, 'Earliest Heart and

Blood Discovered (2014): https://www.nhm.ac.uk/discover/news/2014/april/earliest-heart-blood-discovered.html

FURTHER READING

Stephen Baxter, *Revolutions of the Earth: James Hutton and the True Age of the World* (London: Phoenix, 2004).

Marcia Bjornerud, *Timefulness: How Thinking Like a Geologist Can Help Save the World* (Princeton, NJ: Princeton University Press, 2018).

Richard Fortey, *The Hidden Landscape: A Journey into the Geological Past* (London: The Bodley Head, 2010).

Gabriel Gohau, rev. and trans. Albert V. Carozzi and Marguerite Carozzi, *A History of Geology* (New Brunswick, NJ: Rutgers University Press, 1990).

Stephen Jay Gould, *Times Arrow, Times Circle: Myth and Metaphor in the Discovery of Geological Time* (Cambridge, MA: Harvard University Press, 1987).

Jacquetta Hawkes, *A Land* (Boston, MA: Beacon Press, 1991).

Linda VanAller Hernick, *The Gilboa Fossils* (New York: New York State Museum, 2003).

Susan Elizabeth Hough, *Finding Fault in California: An Earthquake Tourist's Guide* (Missoula, MO: Mountain Press, 2004).

Susan Elizabeth Hough, *Predicting the Unpredictable: The Tumultuous Science of Earthquake Prediction* (Princeton, NJ: Princeton University Press, 2010).

Elizabeth Kolbert, *The Sixth Extinction: An Unnatural History* (London: Bloomsbury, 2014).

Paul Lyell, *The Abyss of Time: A Study in Geological Time and Earth's History*, (Edinburgh: Dunedin Academic Press, 2016).

Charles Lyle, *Principles of Geology*, abridged edn (London: Penguin Classics, 2005).

Donald B. McIntyre and Alan McKirdy, *James Hutton: The Founder of Modern Geology* (Edinburgh: National Museums Scotland, 2012).

Michael McKimm, *Fossil Sunshine* (Tonbridge: Worple Press, 2013).

John McPhee, *Annals of the Former World* (New York: Farrar, Straus and Giroux, 2000).

John Mendum, Jon Merritt and Alan McKirdy, *Northwest Highlands: A Landscape Fashioned by Geology* (Perth: Scottish Natural Heritage, 2001).

Darren Naish and Paul M. Barrett, *Dinosaurs: How They Lived and Evolved* (London: Natural History Museum, 2018).

Ted Nield, *Underlands: A Journey Through Britain's Lost Landscape* (London: Granta, 2014).

Naomi Oreskes (ed.), *Plate Tectonics: An Insiders History of the Modern Theory of the Earth* (Boulder, CO: Westview Press, 2001).

www.nwhgeopark.com

Graham Park, *Introducing Geology: A Guide to the World of Rocks* (Edinburgh: Dunedin Academic Press, 2010).

Martin J. Rudwick, *Earth's Deep History: How It Was Discovered and Why It Matters* (Chicago, IL: University of Chicago Press, 2014).

Richard C. Selley, *The Winelands of Britain: Past, Present & Prospective* (Dorking: Petravin Press, 2008).

Peter Toghill, *The Geology of Britain* (Ramsbury: The Crowood Press, 2002).

David L. Ulin, *The Myth of Solid Ground* (London: Penguin Books, 2005).

Further Reading

Gilbert White, *The Natural History of Selborne* (London: Penguin Classics, 1977).

Simon Winchester, *The Map that Changed the World* (London: Penguin Books, 2002).

Jan Zalasiewicz, *The Earth After Us: What Legacy Will Humans Leave in the Rocks?*, (Oxford: Oxford University Press, 2008).

Jan Zalasiewicz, *The Planet in a Pebble* (Oxford: Oxford University Press, 2012).

ACKNOWLEDGEMENTS

While I was researching this book, many very knowledgeable, generous people gave up their time to talk to me.

For showing me around the Geological Society of London and talking about poetry, thanks to Michael McKimm. Thanks also to Michael and to Bryan Lovell for *Poetry and Geology: A Celebration*.

Jørgen Peder Steffensen kindly showed me the University of Copenhagen's ice core collection and explained the science of glaciology.

Alan McKirdy and Martin Rudwick generously read and commented on sections of this book relating to James Hutton and the history of deep time.

For letting me see inside the world of auction houses, thanks to James Hyslop.

Philip Gibbard, John Marshall and Jan Zalasiewicz were patient and instructive guides to the realm of stratigraphy. Phil was kind enough to share his unpublished research on Giovanni Arduino, and Jan took the time to show me the Jurassic rocks of Leicestershire, answered numerous email queries and shared his writings on Mary Anning and the Comte de Buffon.

Many thanks to Philip Heron and Carolina Lithgow-Bertelloni for helping me understand plate tectonics, and Joan Fryxell for the map and an introduction to the San Andreas fault. Susan Hough kindly drove me around Hollywood on a fault-finding trip. Thanks also to the Kate Scharer and Stan Schwarz of the United States Geological Survey.

Acknowledgements

For taking me on the Chilterns field trip, many thanks to Andrew Farrant and Romaine Graham of the British Geological Survey. Thanks also to Christopher White, the Chief Executive of Denbies, for showing me round the vineyard.

Enzo Morra was invaluable when it came to writing about Campi Flegrei and urban geology in Naples, giving up his time to take me on walking tours of the city, boat trips across the bay and to some wonderful restaurants. Thanks to all the scientists and government employees that I met in Italy. In particular: Carmine Minopoli, Francesca Bianco, Alessio Langella, Gianluca Minin and the Protezione Civile officers in Rome. Thanks also to Enrico Sacchetti for his photographs and translations. In London, Christopher Kilburn kindly answered many questions about his work with volcanoes.

For reflections on Mary Anning, thanks to Natalie Manifold, and for an introduction to fossil preparation, Dan Brownley. More of Dan's prepping can be seen on the Fossil Academy YouTube channel.

Many thanks to Chris Berry who showed me round the National Museum Cardiff, patiently answered my questions (and numerous follow-up questions) on Devonian trees and was generous with the provision of papers and loan of books. Thanks also to Jennifer Clack for talking to me about her research and introducing Boris the Devonian tetrapod.

Writing about dinosaurs would have been far trickier if I hadn't met Michael Leschin of the Bureau of Land Management and Kenneth Carpenter of the Utah State University Eastern Prehistoric Museum. Thanks also to Laverne Antrobus for helping me explore why children love dinosaurs.

For sharing their research into palaeocolour, thanks to Jakob Vinther, Johan Lindgren and Maria McNamara. For introducing me to palaeo art, thanks are due to Robert Nicholls. Bob's work can be viewed and purchased here: http://paleocreations.com

Ruth Siddall first introduced me to the strange world of deep time and geology. I'd like to thank her for some delightful urban geology walks, for reading and commenting on my first urban geology article, and most of all for arranging the walk where I met my husband. You can find out more about urban geology in the UK on the site Ruth helps run: http://londonpavementgeology.co.uk/ (N.B. no longer only covering London.)

For reflections on the Anthropocene, thanks to Kodwo Eshun and The Otolith Group.

Researching nuclear waste in Finland and France was greatly aided by conversations and interviews with Florian Blanquer, Jean-Noël Dumont, Jussi Heinonen, Tina Jalonen, Benoît Jaquet, Anne Kontula, Peter Roche, Rebecca Tadesse, Gerry Thomas. Thanks to Mathieu Saint-Louis for taking me around Cigeo and Pasi Tuohimaa for the trip to Onkalo and the tour of snowy Rauma.

For help locating and understanding the ancient rocks of Scotland, many thanks to Pete Harrison of the North West Highlands Geopark. You can find out more about the Geopark here: https://www.nwhgeopark.com/

Thank you to the magazine editors who commissioned me to write about deep time and geology, and whose support and editorial guidance was invaluable, including Jonathan Beckman, Emma Duncan, Chrissie Giles, João Medeiros, Greg Williams and Simon Willis. Thanks especially to Tim de Lisle who encouraged me to report on urban geology and the Anthropocene – the first deep time writing I produced.

I was fortunate enough to receive travel funding from the University of Hertfordshire, where I teach Creative Writing. Thanks particularly to Rowland Hughes for assistance with funding applications.

My agent, Lisa Baker, first encouraged me to turn an idea into a book. For that and for all her help, support and enthusiasm during the writing and publication process, I'm ever grateful.

Thank you to my editor, Ed Lake, for championing this book. His careful readings, edits and suggestions immeasurably improved my original manuscript. Thanks to everyone at Profile especially

Acknowledgements

managing editor Penny Daniel, Matthew Taylor for the careful copy-edit, Valentina Zanca and Bill Johncocks, and Peter Dyer for the beautiful cover.

Thank you to all my friends and family who have listened to me talking about rocks for the last few years. Particular thanks to Emily Bick for sharing her thoughts on nuclear waste, deep time and music; Amber Dowell for many excellent rocky discussions in Wales, the Lake District and other beautiful places; Travis Elborough for writing and publishing advice; Douglas, Natalie, Henry and Rory Gordon for taking me to the Natural History Museum; Richard Paul for sending useful articles on geology and engineering; Mike Smith for dinosaur parties and The Great Dying; the Walking Group for trips to the Ridgeway and elsewhere.

Just for being herself, thanks to Greta Gordon, who makes an appearance in this book as a small heartbeat and considerately delayed her eventual arrival until six hours after the final edit was complete.

Finally, I want to thank my husband, Jonathan Paul, without whose expertise and support this book could never have been written. Thank you for being a patient advisor on matters scientific, a sharp-eyed reader and an endlessly wonderful travelling companion.

INDEX

Note: *Italic* entries are either titles of books, films or artwork or the names of extinct genera

Index

Index

Index